RF MEMS Circuit Design for Wireless Communications

For a listing of recent titles in the *Artech House Microelectromechanical Systems (MEMS) Series,* turn to the back of this book.

RF MEMS Circuit Design for Wireless Communications

Héctor J. De Los Santos

Artech House
Boston • London
www.artechhouse.com

Library of Congress Cataloging-in-Publication Data
De Los Santos, Héctor J.
    RF MEMS circuit design for wireless communications/Héctor J. De Los Santos.
        p. cm.—(Artech House microelectromechanical systems library)
    Includes bibliographical references and index.
    ISBN 1-58053-329-9 (alk. paper)
    1. Wireless communication systems—Equipment and supplies.   2. Radio circuits.
    3. Microelectromechanical systems.
        I. Title.   II. Series.
    TK5103.2.S26   2002
    621.382—dc21   2002016428

British Library Cataloguing in Publication Data
De Los Santos, Héctor J.
    RF MEMS circuit design for wireless communications. — (Artech House
    microelectromechanical systems series)
    1. Electronic circuit design.   2. Radio frequency.   3. Microelectromechanical systems.
    I. Title
    621.3'815
    ISBN 1-58053-329-9

Cover design by Igor Valdman

International Standard Book Number: 1-58053-329-9
Library of Congress Catalog Card Number: 2002016428

10 9 8 7 6 5 4 3 2 1

Este libro lo dedico a mis queridos padres y a mis queridos, Violeta, Mara, Hectorcito, y Joseph.

Y sabemos que a los que aman a Dios todos los cosas las ayudan a bien, esto es, a los que conforme a su propósito son llamados.

Romanos 8:28

Contents

Preface

This book examines the recent progress made in the emerging field of microelectromechanical systems (MEMS) technology in the context of its imminent insertion and deployment in radio frequency (RF) and microwave wireless applications. In particular, as the potential of RF MEMS to enable the implementation of sophisticated, yet low-power, portable appliances that will fuel the upcoming wireless revolution gains wide recognition, it is imperative that the knowledge base required to quickly adopt and gainfully exploit this technology be readily available. In addition, the material presented herein will aid researchers in mapping out the terrain and identifying new research directions in RF MEMS. Accordingly, this book goes beyond an introduction to MEMS for RF and microwaves [which was the theme of our previous book *Introduction to Microelectromechanical (MEM) Microwave Systems* (Artech House, 1999)] and provides a thorough examination of RF MEMS devices, models, and circuits that are amenable for exploitation in RF/microwave wireless circuit design.

This book, which assumes basic, B.S.-level preparation in physics or electrical engineering, is intended for senior undergraduate or beginning graduate students, practicing RF and microwave engineers, and MEMS device researchers who are already familiar with the fundamentals of both RF MEMS and traditional RF and microwave circuit design.

Chapter 1 of *RF MEMS Circuit Design for Wireless Communications* starts by clearly stating the ubiquitous wireless communications problem, in particular, as it relates to the technical challenges in meeting the extreme levels of appliance functionality (in the context of low power consumption) demanded by consumers in their quest for connectivity at home, while on the move, or on a global basis. The chapter continues with a review of the wireless standards, systems, and traditional architectures, as well as their limitations, which, in turn, are imposed by those of the conventional RF technologies they utilize. Finally, it posits the real prospect of RF MEMS as the technology that can overcome these limitations and thus enable the ubiquitous connectivity paradigm.

Chapter 2 provides a review of those salient points in the discipline of RF circuit design that are key to its successful practice and are intimately related to the successful exploitation of RF MEMS devices in circuits. In particular, the subjects of skin effect, the performance of transmission lines on thin substrates, self-resonance frequency, quality factor, moding (packaging), DC biasing, and impedance mismatch are discussed.

Chapter 3 provides an in-depth examination of the arsenal of MEMS-based devices on which RF MEMS circuit design will be predicated—namely, capacitors, inductors, varactors, switches, and resonators, including pertinent information on their operation, models, and fabrication. The chapter concludes with a discussion of a paradigm for modeling RF MEMS devices using three-dimensional (3-D) mechanical and full-wave electromagnetic tools, in the context of self-consistent mechanical and microwave design.

Chapter 4, via a mostly qualitative treatment, provides a sample of the many novel devices and circuits that have been enabled by exploiting the degrees of design freedom afforded by RF MEMS fabrication techniques—in particular, reconfigurable circuit elements, such as inductors, capacitors, LC resonators, and distributed matching networks; reconfigurable circuits, such as stub-tuners, filters, oscillator tuning systems, RF front-ends, and phase shifters; and reconfigurable antennas, such as tunable dipole and tunable microstrip patch-array antennas.

Chapter 5 integrates all the material presented up to that point as it examines perhaps the most important RF MEMS circuits—namely, phase shifters, filters, and oscillators—via a number of case studies. These include X-band and Ka-band phase shifters for phased arrays and radar applications, film bulk acoustic (FBAR) filters for PCS communications, MEM resonator-based filters, micromachined cavity- and MEM resonator-based oscillators, and a MEM varactor-based voltage-controlled oscillator (VCO). Each case study provides an examination of the particular circuit in terms of its specification and topology, its circuit design and implementation, its circuit packaging and performance, and an epilogue on lessons learned.

Acknowledgments

The author thanks the management of Coventor, in particular, Mr. R. Richards, Mr. J. Hilbert, and Mr. G. Harder, for allowing him to undertake this project. Special thanks are due to the many colleagues who responded rather promptly to his request for original artwork: Dr. A. Muller (IMT-Bucharest), Dr. Yanling Sun (Agere Systems), Dr. J.-B. Yoon (KAIST), Dr. G. W. Dahlmann (Imperial College, London), Drs. R. E. Mihailovich, J. DeNatale, and Y.-H. Kao Lin (Rockwell Scientific Corporation), Mr. M. Stickel and Prof. G. V. Eleftheriades (University of Toronto), Mr. H. Maekoba (Coventor), Dr. F. De Flaviis and Mr. J. Qian (University of California, Irvine), Prof. T. Weller and Mr. T. Ketterl (University of South Florida), Dr. Katia Grenier (Agere Systems), Dr. Y. Kwon (Seoul National University), and Mr. J. Kiihamäki (VTT Electronics).

Special thanks go also to Dr. C. M. Jackson (Ditrans Corporation) for loaning to the author part of his personal technical library collection. Mr J. Repke (Coventor) is thanked for providing useful links to wireless standards.

The author also gratefully acknowledges the cooperation of Ms. J. Hansson and Mr. W. J. Hagen, both of the IEEE Intellectual Property Rights Department, for expediting the granting of a number of permission requests; of Ms. M. Carlier, Mr. S. Tronchon, and Mr. K. Heinz Rosenbrock, all of the European Telecommunications Standards Institute (ETSI), for their assistance in obtaining the permission to reprint excerpts of the GSM standard; and of Ms. A. Essenpreis of the Rights and Permissions

Department, Springer-Verlag, for her assistance in obtaining various permission requests.

The unknown reviewer is thanked for providing useful suggestions on manuscript content and organization. The assistance of the staff at Artech House is gratefully acknowledged, in particular, Mr. Mark Walsh, senior acquisitions editor, for facilitating the opportunity to work on this project, Ms. Barbara Lovenvirth, assistant editor, for her assistance throughout manuscript development, and Ms. Judi Stone, executive editor, for her assistance with the artwork during the production stage. Finally, the author gratefully acknowledges the unfailing and generous assistance of his wife, Violeta, in cutting and pasting artwork throughout the preparation the manuscript.

1

Wireless Systems—A Circuits Perspective

1.1 Introduction

Consumer exigency for ubiquitous connectivity is widely recognized as the demand whose fulfillment will unleash the next industrial revolution beginning in the first decade of the twenty-first century [1]. Such a revolution will be predicated upon the promise to endow these consumers with the ability to achieve universal access to information. The consumers demanding this connectivity, as well as their information needs, are rather diverse. On the one hand, there are individuals, who exploit wireless access for such things as location determination, conversation, personal information management (e.g., calendar of appointments, contact list, address book), checking bank balances, booking movie tickets, finding out about the weather, and money management. On the other, there are businesses, whose information needs may include fleet location, events and status notification, information management, scheduling and dispatch, real-time inventory control, and order and resource management.

Until recently, it was straightforward to associate a single wireless appliance with each one of the various types of information sources (see Figure 1.1). For instance, cell phones were associated with voice, digital cameras with video, laptop computers with broadband data, pagers with messaging, global positioning receivers (GPS) with navigation, and Web appliances with the Internet. The evolution in wireless standards elicited by the growth in

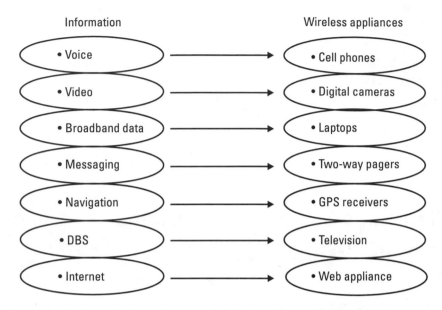

Figure 1.1 Traditional information source/wireless appliance relationship.

consumer demands, however, indicates that expectations from these wireless appliances are getting more and more exacting (see Table 1.1). For example, while the appliances of the first-generation (1G) provided single-band analog cellular connectivity capabilities, those of the second generation (2G) had to provide dual-mode, dual-band digital voice plus data, and now those of the third (3G) and fourth (4G) generations have to provide multimode (i.e., ana-log/digital), multiband (i.e., various frequencies), and multistandard per-formance capabilities. (Various standards include Global System for Mobile Communications (GSM)—a leading digital cellular system that allows eight simultaneous calls on the same radio frequency; Digital European Cordless Telecommunications (DECT)—a system for the transmission of integrated voice and data in the range of 1.88 to 1.9 GHz; cellular digital packet data (CDPD)—a data transmission technology that uses unused cellular channels to transmit data in packets in the range of 800 to 900 MHz; General Packet Radio Service (GPRS)—a standard for wireless communications that runs at 150 Kbps; and code division multiple access (CDMA)—a North American standard for wireless communications that uses spread-spectrum technology to encode each channel with a pseudo-random digital sequence.) The key question then becomes: Will it be possible to realize the wireless appliances

Table 1.1

Wireless Standards—The Evolution Blueprint

1G	2G	3G	4G
Analog cellular (single band)	Digital (dual-mode, dual-band)	Mulitmode, multiband software-defined radio	Multistandard + multiband
Voice telecom only	Voice + data telecom	New services market beyond traditional telecom: higher speed data, improved voice, multimedia mobility	
Macro cell only	Macro/micro/pico cell	Data networks, Internet, VPN, WINternet	
Outdoor coverage	Seamless indoor/outdoor coverage		
Distinct from PSTN	Complementary to fixed PSTN		
Business customer focus	Business + consumer	Total communications subscriber: virtual personal networking	

Source: http://www.uwcc.org.

embodying this convergence of functions and interoperability (Figure 1.2) given the power and bandwidth limitations imposed by conventional RF circuit technology, in the context of ubiquitous connectivity? With this question in mind, we now examine the spheres of influence in which these wireless appliances function, as well as pertinent technical issues, the challenges to enabling power/bandwidth-efficient wireless appliances, and the potential of MEMS technology to enable wireless appliances capable of fulfilling the ubiquitous connectivity vision.

1.2 Spheres of Wireless Activity—Technical Issues

In order to achieve this overarching goal of ubiquitous connectivity by way of all-encompassing and interoperable wireless appliances, it will be necessary to enable seamless, efficient, secure, and cost-effective connectivity for

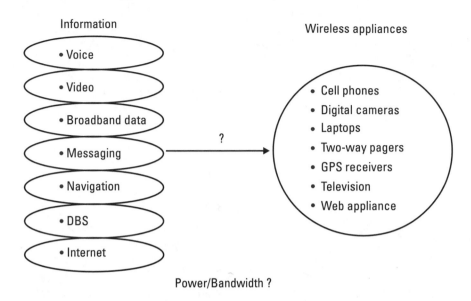

Figure 1.2 Evolution towards total convergence and interoperability wireless appliances.

information appliances operating within and among the various spheres of consumer activity (Figure 1.3): (1) the home and the office, (2) the ground fixed/mobile platform, and (3) the space platform.

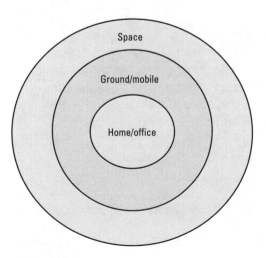

Figure 1.3 Spheres of wireless activity.

With mobility and portability as the common themes, the plans for these 3G mobile wireless telecommunications services call for supporting mobile and fixed users who employ a wide variety of devices, including small pocket terminals, handheld telephones, laptop computers, and fixed-receiver appliances operating at frequencies that take advantage of the excellent properties of radio waves below 3 GHz [2].

Complete success in bringing this vision to fruition, however, may well depend on our ability to harness two scarce currencies, namely, power and bandwidth. Power is essential due to the overt conflict between increased levels of sophistication and functionality demanded of the mobile information appliances, and the limited battery power available [3]. Bandwidth, on the other hand, is crucial because of the large population of wireless devices already operating below 3 GHz. We will show that microelectromechanical systems (MEMS) technology, as applied to these information appliances, is poised as the source capable of generously supplying these two key resources. Thus, it is the main goal of this book to provide the background necessary to exploit MEMS technology in the design of the RF circuits that will enable the fulfillment of this vision in the context of a wireless paradigm. We begin this exposition with an examination of the various realms of wireless activity and their respective information appliances and performance needs. Then we introduce the fundamental circuit and systems elements whose performance level is key to determining the success of the wireless paradigm. Finally, we point out the intrinsic features of MEMS technology that make it the ideal candidate to enable the realization of these circuit and system functions, together with a number of specific early examples that validate our expectations of the power of MEMS to enable the wireless vision.

1.2.1 The Home and the Office

The advent and perfecting of the microprocessor that began in the 1970s and 1980s enabled the conception of ever more powerful and intelligent stand-alone home appliances—for example, television sets, microwave ovens, stereo systems, telephones, lighting control, surveillance cameras, climate-control systems, and the personal computer (PC). The office environment, on the other hand, motivated by the pursuit of increases in productivity and cost efficiency, saw the massive deployment in the early 1990s of wired networks to link office appliances—for example, PCs, servers, workstations, printers, and copiers. Finally, with the explosion in the late 1990s of consumer appetite for access to information, brought about by home-PC-enabled Internet access, the conception and deployment of products and services revolving around the ubiquitous retrieval, processing, and transport of information has made the home an important part of the global communications grid.

Thus, the home market, which lagged behind the office in adopting local area networks, is now the battlefield of competing networking technologies that aim at enabling a new level of connectivity by exploiting emerging networking-ready appliances and the infrastructure already present in the home (e.g., voice-grade telephone wiring, twisted pairs, power lines, and, increasingly, wireless links). Wireless short-range links are particularly attractive because, in addition to being a convenient medium for voice, video, and data transport, they can provide inexpensive networking solutions in the home or small home-office environment [4]. In fact, an examination of the evolution in home-networked households in the United States reveals a steady increase in the migration from wired networks, based on phone and power lines, to wireless-based networks (Figure 1.4).

Anticipating the potential home wireless networking market, various standards are under development: (1) Bluetooth—a short-range radio technology that supports only voice and data, and that is aimed at simplifying communications among networked wireless appliances and other computers, and (2) HomeRF—a short-range radio technology that supports computer/peripheral networking and wireless Internet access. Both operate at 2.4 GHz [1, 2].

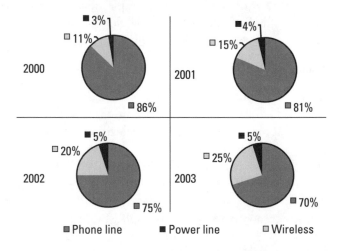

Figure 1.4 Distribution of interconnection media usage for home networks in U.S. households. (*After:* [4].)

Although electronic appliances in this sphere are connected to the power grid (i.e., they are stationary), the issues of power consumption and bandwidth limitations are still critical. The following technical issues must be dealt with to successfully implement these standards: (1) reducing home load to the utilities power grid, due to the great number of power-consuming electronic appliances, which in areas like California is outstripping the available capacity; and (2) reducing the intrinsic signal loss of frequency-selection passive circuits, due to the characteristically low radiated signal powers typical of indoor environments.

1.2.2 The Ground Fixed/Mobile Platform

Since consumers are no longer satisfied with home-PC-based Internet access, demand for ubiquitous wireless access to information while on the move has elicited a plethora of new products and services [e.g., location-aware navigation guides, finance applications, wireless ID cards, freight and fleet management, telemetry, smart-phones, personal digital assistants (PDAs), and laptop computers] that cater to these demands. In this context, because the ability to move seamlessly between independently operated Internet Protocol (IP) networks [3, 5] (e.g., between various countries) will be extremely important, appliances must be equipped to operate over a wide variety of access and network technologies and standards [5], such as GSM, DECT, CDPD, GPRS, and CDMA. Thus, unlike conventional wireless devices, next-generation information appliances will have to include multimode, multiband capabilities [5], along with the concomitant processing overhead associated with function management. In order to successfully integrate these capabilities, two key technical issues must be dealt with: (1) lowering the required power consumption, given the already limited (battery) power source available, and (2) minimizing the mass (weight) of the appliances so as to not hinder their portability.

1.2.3 The Space Platform

The last sphere of activity that enables the global ubiquitous communications vision is the space segment. Indeed, to achieve worldwide access to, and distribution of, the large volumes of integrated voice, video, data, and multimedia information generated in the home/office and ground fixed/mobile realms, space-based platforms (satellites) must be tapped. Unfortunately, the demands for higher capacity and flexibility that this vision imposes on conventional satellites is in direct conflict with the inherent limitations posed by the prohibitive mass and power consumption needed to satisfy these requirements [6]. In particular, meeting these requirements necessitates satellite architectures capable of multiuser, multidata rate, and multilocation links (while exhibiting, for some applications, very small latency). These capabilities, in turn, dictate the utilization of low-loss/low-power-consumption switch matrices and phased-array (electronically steerable) antennas. Thus, to enable this segment of the wireless communications grid, it will be necessary to deal with two key technical issues: (1) the limited on-board power source available, and (2) the conflict posed by the direct relation between capacity and functionality on the one hand, and power consumption and mass on the other.

1.3   Wireless Standards, Systems, and Architectures

1.3.1   Wireless Standards

The implementation of wireless connectivity is predicated upon the definition of so-called wireless standards, of which GSM, DECT, CDPD, GPRS, and CDMA are examples [7, 8]. Each of these standards embodies the precise set of parameters that dictate the architecture and software design of wireless systems operating under the standard to effect intelligible communication with other systems also operating within the standard. The parameters defining a given standard may be classified into those that pertain to the air interface (or front-end) of the system, and those that pertain to the subsequent signal processing (or baseband) part. Among the parameters defining the former, we have multiple access, frequency band, RF channel bandwidth, and duplex method; among those defining the latter, we have modulation, forward and reverse channel data rate, channel coding, interleaving, bit period, and spectral efficiency. Of particular interest to us in this book are the RF-related air interface parameters, shown in Tables 1.2 to 1.4 [7, 8] for a representative sample of standards, as they dictate the nature of the transceiver architectures implementing them.

1.3.2   Conceptual Wireless Systems

As indicated in the previous section, at the core, wireless information appliances may be conceptualized as shown in Figure 1.5. They comprise an antenna and front-end and baseband sections. The antenna effects either detection or emission of electromagnetic signals; the front-end selects, amplifies, and down-converts the received signal, or up-converts, amplifies, and filters the signals to be transmitted. Thus, the antenna and front-end embody

Table 1.2

RF-Related Air Interface Parameters for Analog Cellular Standards

Parameters	Analog Cellular Wireless Systems	
	AMPS	ETACS
Multiple access	FDMA	FDMA
Frequency band (MHz)	824–849/869–894	872–904/917–950
RF channel bandwidth	30 KHz	25 KHz
Duplex method	FDD	FDD

Table 1.3

RF-Related Air Interface Parameters for Digital Cordless Standards

Parameters	Digital Cordless Systems	
	DECT	PHS
Multiple access	FHMA/TDMA	TDMA
Frequency band (MHz)	1,880–1,900	1,895–1,918
RF channel bandwidth	1.728 MHz	300 KHz
Duplex method	TDD	TDD

Table 1.4

RF-Related Air Interface Parameters for 2G Digital Cellular Standards

Parameters	Digital Cellular Systems	
	IS-54	GSM
Multiple access	TDMA	TDMA/FDMA
Frequency band (MHz)	824–849/869–894	890–915/935–960
RF channel bandwidth	30 KHz	200 KHz
Duplex method	FDD	FDD

the air interface (i.e., that part of the system responsible for its wireless ability). The baseband section, on the other hand, effects demodulation and processing of the received carrier signal to extract its information, or modulation of the carrier signal to be transmitted with the information to be communicated. Thus, the baseband section personalizes or defines the function performed by the wireless system (e.g., it makes it a telephone or a pager).

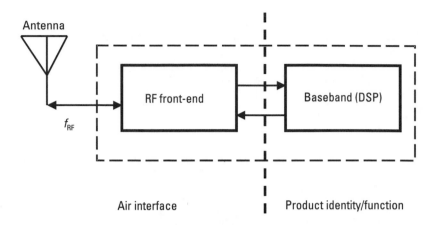

Figure 1.5 Conceptualized wireless information appliance.

1.3.3 Wireless Transceiver Architectures

In this section, we present simplified block diagrams of wireless transceiver architectures that implement the PHS, GSM, and DECT standards presented above using conventional RF technology (Figures 1.6 to 1.9) [9–13]. These figures expose the various solutions and compromises forced by technology limitations, which dictate the architecture's partitioning between integrated and discrete components.

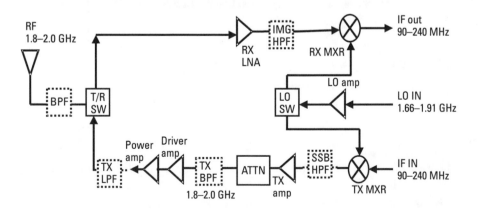

Figure 1.6 Simplified PHS transceiver architecture; components inside dashed boxes are located off-chip.

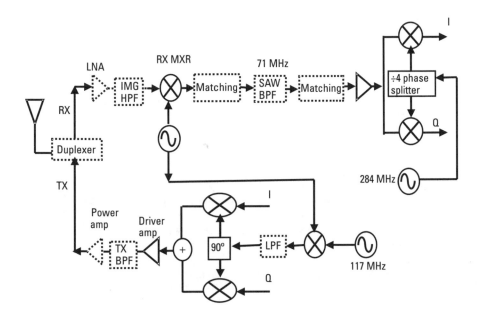

Figure 1.7 Simplified GSM transceiver architecture.

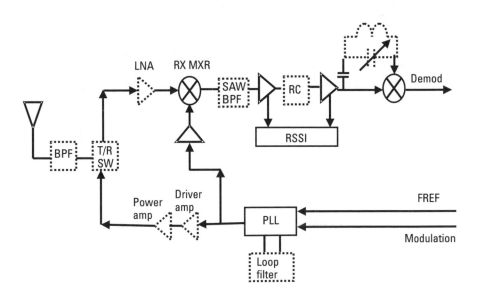

Figure 1.8 Simplified DECT transceiver architecture.

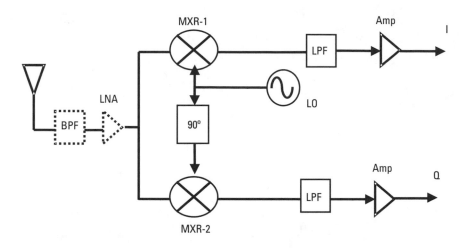

Figure 1.9 Simplified direct conversion receiver architecture.

1.4   Power- and Bandwidth-Efficient Wireless Systems— Challenges

An examination of the transceiver architectures presented above reveals that, in their implementation, several scenarios are encountered.

The depictions in Figure 1.10(a, b) assume omnidirectional antennas. In these cases low-loss/parasitic-free passive elements (e.g., transmission lines, inductors, capacitors, varactors, switches, and resonators) for minimum insertion loss matching networks, tunability, and filtering are imperative [6]. Because of congested spectrum and communication activity in certain environments, it may be necessary to endow wireless appliances with the ability to spatially filter the received signals by nulling undesired interference, to automatically account for poor propagation characteristics, and to maintain the link while on the move [3, 5]. In this case the ability to incorporate phase shifting, summing, and weighting, while introducing minimum loss, is invaluable [Figure 1.10(c)].

On the other hand, since a long battery life is highly desirable, it is clear that increasing the dc-to-RF conversion efficiency by minimizing insertion loss during transmission mode must also be addressed. For example, overcoming losses in the antenna, filtering, and switching circuits would drive the efficiency from 25% to 40% [12].

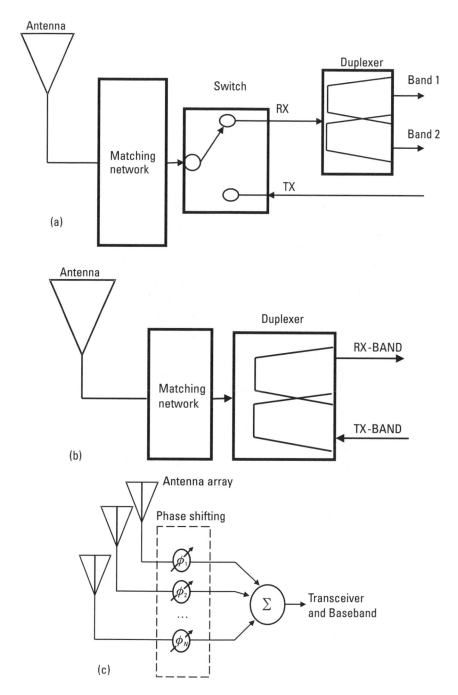

Figure 1.10 Front-end variations: (a) antenna-matching network-T/R switch-duplexer; (b) antenna-matching network-duplexer; (c) smart antenna-transceiver. (After: [1].)

The next block met by the received signal is the receiver (Figure 1.11). Here, the signal is amplified by a low noise amplifier (LNA), filtered, and applied to a mixer, which is driven by a low-phase noise oscillator.

Low-loss performance of the passives is crucial for minimizing transceiver power dissipation. In particular, the quality factor of inductors—either to increase the gain of LNAs while keeping their current consumption down, or to improve the phase noise of oscillators—must be as high as possible [13]. On the transmit mode, it is well documented [14] that the power amplifier dominates power consumption. In fact, lossy substrates give rise to high-loss inductors in matching networks, which in turn results in reduced output power and efficiency [14].

The root causes limiting the ultimate power/bandwidth performance of wireless appliances in all spheres may be traced to substrate parasitics as embodied in its resistivity and dispersion [6]. Low resistivity, in the case of silicon wafers, is responsible for low quality factor, which affects inductors, or high insertion loss, which affects transmission lines [6]. Similarly, in the case of control and tunable elements (e.g., switches and varactors), it is the nature of the semiconductor wafer process that gives rise to high insertion loss and bandwidth-limiting reactive coupling.

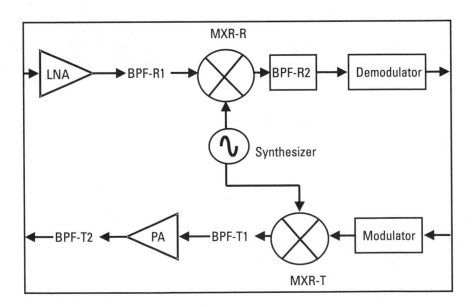

Figure 1.11 Simplified transceiver diagram. (After: [1].)

1.5 MEMS-Based Wireless Appliances Enable Ubiquitous Connectivity

MEMS technology, with its versatility to integrate both electronic (2-D) and micromechanical (3-D) devices, is poised as the rich source capable of generously supplying the two key resources on which the wireless paradigm hinges, namely, low power consumption and bandwidth. Indeed, by judiciously combining its surface micromachining, bulk micromachining, and LIGA fabrication techniques, it is entirely possible to realize virtually parasitic-free RF components [6]. Figure 1.12 shows the arsenal of high-quality components enabled by MEMS.

A recent summary of the performance of state-of-the-art RF MEMS components bears out our optimism [15]. In particular, bulk-etched 1nH inductors exhibited measured quality factors Q ranging between 6 and 28 at frequencies between 6 and 18 GHz, while surface micromachined 2.3-nH inductors exhibited a Q of 25 at 8.4 GHz. Surface micromachined 2.05-pF varactors exhibited a Q of 20 at 1 GHz, for a capacitance tuning range of 1.5:1 over a 0 to 4V tuning voltage at a self-resonance frequency of 5 GHz. MEM switches have exhibited a series resistance less than 1 Ohm, insertion loss less than 0.1 dB at 1 GHz, isolation greater than 40 dB at 1 GHz, third-

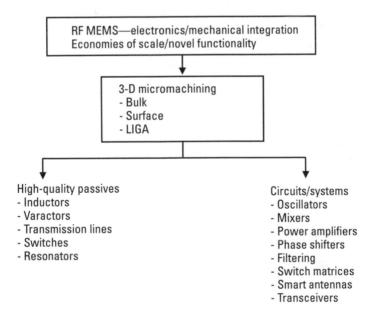

Figure 1.12 MEMS-enabled RF components. (*After:* [1].)

order intercept point (IP3) greater than +66 dBm, actuation voltage between 3 and 30V, and control current less than 10 uA. Micromachined cavity resonators have demonstrated Qs of 500 at 10 GHz, only 3.8% lower than the unloaded Q obtained from a rectangular cavity of identical dimensions. Microelectromechanical resonators, on their part, have exhibited Qs of 7,450 at 92.5 MHz, while film bulk acoustic resonators (FBARs) have exhibited Qs of over 1,000 at resonance frequencies between 1.5 and 7.5 GHz. Finally, it must be noted that the improvement obtained by bulk etching the substrate to eliminate the parasitics of transmission lines has been remarkable. For instance, insertion loss improvements of 7 dB at 7 GHz and 20 dB at 20 GHz have been attained.

In addition to the above results, results on individual components are beginning to appear in the literature, which demonstrate successful production-grade RF MEMS circuits. For instance, Agilent's 1,900-MHz FBAR duplexer for Personal Communications Services (PCS) handsets is set for high-volume production. The duplexer enables a size 5X reduction with respect to its ceramic counterpart at comparable performance.

While there is much excitement as we witness the dawn of the revolution brought about by the potentialities of MEMS in wireless applications, we must keep in mind that it took several decades for the previous revolution—brought about by integrated circuit technology—to reach its full potential. Similarly, we must grow through the pains associated with increasing integration levels and demonstrating adequate reliability. In these contexts, much work remains to be done in the development of modeling and circuit-/system-level design methodologies for the multiphysics-multidomain devices characteristic of MEMS [6], and their proper metrology and reliability assessments.

1.6 Summary

This chapter dealt with the imminent revolution in wireless communications expected to be triggered by the growth in consumer demand for ubiquitous wireless access to information. The ability to successfully meet such a demand by systems based on conventional RF technologies is questionable, given the exacting requirements for power and bandwidth efficiency demanded by these sophisticated systems. Upon reviewing the factors that shape the engineering of wireless systems—namely, their spheres of operation, standards, and architectures—we turned to examining the nature of the limitations imposed by conventional RF technologies. Indeed, these limitations were traced to two: substrate parasitics and device fabrication process. Against this backdrop, MEMS was presented as a powerful technology capable of enabling devices to overcome these limitations. Examples of MEMS technology's vast arsenal of fabrication processes and device techniques were then given to substantiate the belief that this technology offers a rich resource to overcome the key factors limiting ubiquitous wireless connectivity—namely, power and bandwidth.

References

[1] De Los Santos, H. J., "MEMS—A Wireless Vision," *2001 International MEMS Workshop*, Singapore, July 4–6, 2001.

[2] "The Potential for Accommodating Third Generation Mobile Systems in the 1710–1850 MHz Band: Federal Operations, Relocation Costs, and Operational Impacts," Final Report, http://www.ntia.doc.gov/ntiahome/threeg/33001/3g33001.pdf, March 2001.

[3] Parrish, R. R., "Mobility and the Internet," *IEEE Potentials Magazine*, April/May 1998, pp. 8–10.

[4] Dutta-Roy, A., "Networks for Homes," *IEEE Spectrum*, Vol. 36, 1999, pp. 26–33.

[5] Fasbender, A., et al., "Any Network, Any Terminal, Anywhere," *IEEE Personal Communications Magazine*, April 1999, pp. 22–30.

[6] De Los Santos, H. J., *Introduction to Microelectromechanical (MEM) Microwave Systems*, Norwood, MA: Artech House, 1999.

[7] Rappaport, T. S., *Wireless Communications: Principles and Practice*, Englewood Cliffs, NJ: Prentice Hall, 1996.

[8] Momtahan, O., and H. Hashemi, "A Comparative Evaluation of DECT, PACS, and PHS Standards for Wireless Local Loop Applications," *IEEE Communications Magazine*, May 2001, pp. 156–162.

[9] McGrath, F., et al., "A 1.9 GHz GaAs Chip Set for the Personal Handyphone System," *IEEE Trans. Microwave Theory Tech.*, Vol. 43, 1995, pp. 1733–1744.

[10] Stetzler, T. D., et al., "A 2.7–4.5 V Single Chip GSM Transceiver RF Integrated Circuit," *IEEE J. Solid-State Circuits*, Vol. 30, No. 12, December 1995, pp. 1421–1429.

[11] Heinen, S., "Integrated Transceivers for Digital Cordless Applications," *2000 IEEE Bipolar Circuits and Technology Meeting*, Minneapolis, MN, September 24–26, 2000 pp. 44–51.

[12] Baltus, P. G. M., and R. Dekker, "Optimizing RF Front Ends for Low Power," *Proc. IEEE*, Vol. 88, October 2000, pp. 1546–1559.

[13] Abidi, A. A., G. J. Pottie, and W. Keiser, "Power-Conscious Design of Wireless Circuits and Systems," *Proc. IEEE*, Vol. 88, 2000, pp. 1528–1545.

[14] Gupta, R., B. M. Ballweber, and D. J. Allstot, "Design and Optimization of CMOS RF Power Amplifiers," *IEEE J. Solid-State Circuits*, Vol. 35, 2001, pp. 166–175.

[15] Richards, R. J., and H. J. De Los Santos, "MEMS for RF/Wireless Applications: The Next Wave," *Microwave J.*, March 2001.

2

Elements of RF Circuit Design

2.1 Introduction

The design of RF MEMS circuits for wireless applications is predicated upon the well-established principles of RF and microwave electronics, as well as on the novelty of RF MEMS devices. Although, for design purposes, RF MEMS devices are dealt with via the abstraction of design-oriented circuit models, their dominant 3-D nature demands a special awareness pertaining to certain physical aspects. In this chapter, we briefly address some of these aspects: the skin effect, microstrip and coplanar waveguide transmission lines, self-resonance frequency, quality factor, moding (packaging), dc biasing, and impedance mismatch.

2.2 Physical Aspects of RF Circuit Design

Ideally, RF and microwave circuits are comprised of interconnections of well-demarcated components. These components include lumped passive elements [1] (such as resistors, capacitors, and inductors), distributed elements [2] (such as microstrip, coplanar waveguide, or rectangular waveguide), and active elements [3, 4] [such as field-effect transistors (FETs) or bipolar transistors]. Often, control elements to effect signal switching and routing [2] (such as pin diode switches or FET switches) are also utilized. Configuring circuit models of these elements according to a circuit topology that defines the desired function, along with the help of a computer-aided design (CAD) tool, one eventually arrives at a circuit whose performance meets specifications and is ready for the next steps of fabrication and testing.

Unfortunately, this simplistic vision of RF and microwave circuit design often becomes blurred when test results are obtained that differ drastically from the beautiful simulation results. The reasons for this disparity may normally be traced to one of the following:

- The frequency of operation is such that the circuit elements display complex behavior, not represented by the pure element definitions utilized during the design.

- The circuit layout includes coupling paths not accounted for in the design.

- The ratio of the transverse dimensions of transmission lines to wavelength are nonnegligible—thus, additional unwanted energy storage modes become available.

- The package that houses the circuit becomes an energy storage cavity, thus absorbing some of the energy propagating through it.

- The (ideally) perfect dc bias source is not adequately decoupled from the circuit.

- The degree of impedance match among interconnected circuits is not good enough, so that large voltage standing wave ratios (VSWR) are present, which give rise to inefficient power transfer and to ripples in the frequency response.

Below we address each of these concepts.

2.2.1 Skin Effect

Skin effect is perhaps the most fundamental physical manifestation of the RF and microwave frequency regime in circuits. In a conductor adjacent to a propagating field, such as a transmission line or the inside walls of a metallic cavity, because the conductor's resistance is actually nonzero, the propagating field does not become zero immediately at the metal interface but penetrates for a short distance into the conductor before becoming zero [4]. As the distance the field penetrates the conductor varies with frequency, it invades the conductor in the region near the surface, thus occupying a skin of conductor volume. When the field propagates within the conductor in this region of nonzero resistance, it incurs dissipation. In quantitative terms, the skin depth is defined as the distance it takes the field to decay exponentially to $e^{-1} = 0.368$, or 36.8% of its value at the air-conductor interface, and is given by [4]

$$\delta = \frac{1}{\sqrt{f\pi\mu\sigma}} \tag{2.1}$$

where f is the signal frequency, μ is the permeability of the medium surrounding the conductor, and σ is the conductivity of the metal making up the conductor. From this equation, it is clear that the skin depth decreases inversely proportional to the square root of the frequency and the conductivity.

An electromagnetic analysis of the skin depth phenomenon [5] leads to its characterization in terms of the so-called internal impedance of the conductor (Figure 2.1), which for unit length and width is defined as

$$Z_S = R_S + X_S \tag{2.2}$$

where the first term, called the surface resistivity, is given by

$$R_S = \frac{1}{\sigma\delta} = \sqrt{\frac{\pi f \mu}{\sigma}} \tag{2.3}$$

The second term is called the internal reactance of the conductor and is given by [5]

$$X_S = \omega L_i = \frac{1}{\sigma\delta} = R_S \tag{2.4}$$

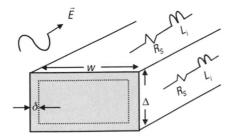

Figure 2.1 Skin effect in metallic conductor of rectangular cross section.

where L_i is referred to as the internal inductance.

Since the skin depth represents the frequency-dependent energy loss due to propagation in the resistive region within a skin depth of the surface, it is important to minimize it. This can be achieved by choosing a conductor metal with high conductivity. For instance, at a frequency of 2 GHz, the skin depth for an aluminum conductor, $\sigma_{Al} = 3.72 \times 10^7 \, (S/m)$, is $\delta_{Al} = 1.85 \, \mu m$; whereas that for copper, $\sigma_{Cu} = 5.8 \times 10^7 \, (S/m)$, is $\delta_{Cu} = 1.47 \, \mu m$. Thus, if the same current, I, is to flow in both conductors, the power dissipated in the copper line will be 80.7% of that dissipated in the aluminum line—a reduction in power loss of almost 20%.

Another important implication of skin depth is that of energy confinement. This issue relates to the minimum conductor thickness of, for example, a planar transmission line, such as microstrip, or of the metallic structures making up RF MEMS devices. Let us consider, for example, the case of copper in the previous paragraph: What happens if the line thickness, Δ, is equal to the skin depth, $1.47 \, \mu m$? We can see in Figure 2.2 that because (by definition) at a skin depth from the surface the incident field has decayed to 36.8% of its maximum value—there being no more conductor after $\Delta_1 = \delta$ that can dissipate the rest of the energy to induce the remaining of the decay down to zero amplitude—the energy escapes. This is like having an energy leakage in the system, since the propagating energy the conductor is supposed to guide or confine has to feed both the energy dissipated in the surface resistance and the energy that escapes. On the other hand, if a conductor of thickness $\Delta_2 = 2\delta$ is utilized, at this distance from the surface the incident field has decayed to 13.5% of its surface value; thus, less energy escapes. When the conductor thickness is chosen as four skin depths, the amplitude at this point reaches a value of 1.8% of its incident value and the

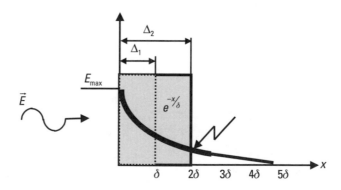

Figure 2.2 Field decay into metallic conductor.

energy that escapes is negligible. Therefore, the minimum conductor thickness should be chosen such that it is approximately equal to, or greater than, four skin depths at the lowest frequency of operation.

2.2.2 Transmission Lines on Thin Substrates

The properties of conventional microstrip lines (i.e., those defined on substrates with thickness in the hundreds of microns) are well known and have been captured in accurate closed-form expressions [6]. These models, however, fail to reproduce the characteristics of microstrip lines defined on substrates with thickness in the neighborhood of just a few microns. The novel thin-film microstrip lines are evoked in a number of contexts, in particular, the development of silicon RF ICs [7]. Indeed, when a metallic ground plane, followed by an insulating dielectric layer, is deposited on top of low-resistivity silicon wafers, the loss properties of the line are decoupled from those of the silicon wafer, thus circumventing its poor microwave properties (Figure 2.3).

Unfortunately, this new dimensional regime does bring to the fore other deleterious phenomena. For instance, the smaller line width dimension

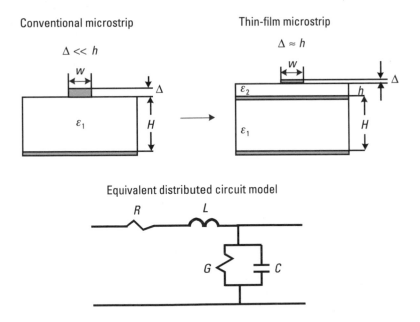

Figure 2.3 Cross sections of conventional and thin-film microstrip lines and equivalent distributed circuit model.

accompanying the thinner substrates results in an increased conductor loss, and the larger conductor thickness-to-substrate ratio demands that the finite metal conductivity and internal inductance effects be included to suitably model the behavior in this regime. Such effects have been synthesized recently in a model advanced by Schneider and Heinrich [8].

2.2.2.1   Thin-Film Microstrip Line Model

In order to derive the thin-film microstrip model, Schneider and Heinrich [8] recognized and exploited the fact that, due to the small dimensions relative to wavelength of these lines, the quasi-TEM approximation could be invoked. Under this approximation [9], the microstrip line is analyzed as if it were suspended in air (i.e., as a TEM line), and then the effect of the substrate is taken into account by adjusting the air-filled-derived values of the propagation constant and characteristic impedance by an effective dielectric constant obtained by performing a static analysis. This resulted in closed-form expressions for frequencies low enough so that dispersion could be ignored. To extend the results to the highest frequencies, at which dielectric dispersion would be important, Schneider and Heinrich incorporated the formulas for dielectric dispersion derived by Kirschning and Jansen [8]. The line was modeled by the conventional series RL-shunt GC, distributed circuit (Figure 2.3), with elements given as follows.

Distributed Capacitance C and Conductance G

$$C = \frac{1}{c \times Z_{L0}\left(w_{eq0}\right)} \times \varepsilon_{r,\text{eff}} \qquad (2.5)$$

where the parameters are given as follows:

- c is the speed of light in free space;
- Z_{L0} is the characteristic impedance of the line assuming zero metal thickness and suspended in air (without substrate), given by

$$Z_{L0} = \frac{\eta_0}{2\pi} \times \ln\left\{ \frac{F_1 \times h}{w} + \sqrt{1 + \left(\frac{2 \times h}{w}\right)^2} \right\} \qquad (2.6)$$

with

$$F_1 = 6 + (2\pi - 6) \times \exp\left\{-\left(30.666 \times \frac{h}{w}\right)^{0.7528}\right\} \tag{2.7}$$

and

$$\eta_0 = \sqrt{\frac{\mu_0}{\varepsilon_0}} \tag{2.8}$$

- w_{eq0} is the equivalent line width, assuming a nonzero thickness, Δ, and suspended in air (without substrate), given by

$$w_{eq0} = w + \frac{\Delta}{\pi} \times \ln\left\{1 + \frac{4 \times \exp(1)}{\frac{\Delta}{h} \times \coth^2\left(\sqrt{6.517 \times \frac{w}{h}}\right)}\right\} \tag{2.9}$$

- $-\varepsilon_{r,eff}$ is the effective relative constant assuming nonzero conductor thickness, given by

$$-\varepsilon_{r,eff} = \varepsilon_{r,eff,0}(w_{eqZ}) \times \left[\frac{Z_{L0}(w_{eq0})}{Z_{L0}(w_{eqZ})}\right]^2 \tag{2.10}$$

with

$$\varepsilon_{r,eff,0}(w) = \frac{\varepsilon_r + 1}{2} + \frac{\varepsilon_r - 1}{2} \times \left(1 + \frac{10 \times h}{w}\right)^{-a \times b} \tag{2.11}$$

$$a = 1 + \frac{1}{49} \times \ln\left(\frac{\left(\frac{w}{h}\right)^4 + \left(\frac{w}{52 \times h}\right)^2}{\left(\frac{w}{h}\right)^4 + 0.432}\right) + \frac{1}{18.7} \tag{2.12}$$

$$\times \ln\left\{1 + \left(\frac{w}{18.1 \times h}\right)^3\right\}$$

$$b = 0.564 \times \left(\frac{\varepsilon_r - 0.9}{\varepsilon_r + 3}\right)^{0.053} \qquad (2.13)$$

$$w_{eqZ} = w + \frac{w_{eq0} - w}{2} \times \left(1 + \frac{1}{\cosh\left(\sqrt{\varepsilon_r - 1}\right)}\right) \qquad (2.14)$$

The corresponding characteristic impedance is given by

$$Z_L = \frac{Z_{L0}\left(w_{eqZ}\right)}{\sqrt{\varepsilon_{r,\text{eff},0}\left(w_{eqZ}\right)}} \qquad (2.15)$$

The distributed conductance G embodies the substrate loss and is given in terms of the dielectric loss tangent by

$$G = \omega \times \frac{\varepsilon_{e,\text{eff}} - 1}{\varepsilon_r - 1} \times \varepsilon_r \times C'_a \times \tan\left(\delta_\varepsilon\right) \qquad (2.16)$$

These formulas for C and G are expected to be valid up to well beyond 100 GHz [8].

Distributed Inductance L and Resistance R

As discussed previously, the frequency-induced skin effect plays an important role in determining the microwave behavior of metallic structures. Such a behavior is a function of the relative dimensions between structural geometries and the skin depth. Thus, while at the lowest frequencies (close to dc) where dimensions are small compared to the skin depth, the skin effect is negligible; at higher frequencies, where dimensions are large compared to the skin depth, the skin effect dominates. Finally, there is an intermediate region, where structure dimensions and skin depth are comparable, that exhibits both behaviors [8]. For modeling purposes, the intermediate frequency range is demarcated by two frequencies: f_0 and f_{se}. The formulas are given as follows:

$$L(f) = L'_a + \frac{L'_i(f_{se})}{1 + \sqrt{\dfrac{f}{f_{se}}}} + \frac{L'_{dc} - L'_a - L'_i(f_{se})}{\sqrt{1 + \left(\dfrac{f}{f_0}\right)^2}} \qquad (2.17)$$

where

$$L'_a = \frac{1}{c^2 C'_a} \tag{2.18}$$

$$L'_i = \frac{R'_{se}(f)}{\omega} \tag{2.19}$$

$$R'_{se} = 2 \times Z_L \times \alpha_c(f) \tag{2.20}$$

$$\alpha_c(f) = 0.1589 \times A \times \frac{R_s(f)}{h \times Z_L} \times \frac{32 - \left(\dfrac{w_{eq0}}{h}\right)^2}{32 + \left(\dfrac{w_{eq0}}{h}\right)^2} \quad \text{for } \frac{w}{h} \leq 1 \tag{2.21}$$

and

$$\alpha_c = 7.0229 \times 10^{-6} \times A \times \frac{R_s(f) \times Z_L \times \varepsilon_{r,\text{eff}}}{h} \times$$

$$\left[\frac{w_{eq0}}{h} + \frac{0.667 \times \dfrac{w_{eq0}}{h}}{\dfrac{w_{eq0}}{h} + 1.444}\right] \quad \text{for } \frac{w}{h} \geq 1 \tag{2.22}$$

with

$$A = 1 + \frac{h}{w_{eq0}} \times \left[1 + \frac{1.25}{\pi} \times \ln\left(\frac{2 \times h}{\Delta}\right)\right] \tag{2.23}$$

and R_s given by (2.3), $R_s(f) = \sqrt{\dfrac{\pi f \mu}{\sigma}}$. The intermediate frequency regime boundaries, f_0 and f_{se}, are given by

$$f_0 = \frac{2 \times R'_{dc,w} \times R'_{dc,g}}{\mu_0 \times \left(R'_{dc,w} + R'_{dc,g}\right)} \tag{2.24}$$

and

$$f_{se} = \frac{1.6 + \dfrac{10 \times \dfrac{\Delta}{w}}{1 + \dfrac{w}{h}}}{\pi \times \mu_0 \times \sigma \times \Delta^2} \qquad (2.25)$$

where

$$R'_{dc,w} = \frac{1}{\sigma \times w \times \Delta} \qquad (2.26)$$

and

$$R'_{dc,g} = \frac{1}{\sigma \times w_g \times \Delta} \qquad (2.27)$$

are the resistances of the line and ground conductors, respectively.

L'_{dc}, the low frequency inductance, embodies the uniformly distributed current in this regime and is given by [10]

$$L'_{dc} = \frac{-\mu_0}{2 \times \pi \Delta^2} \times \left[\frac{1}{w^2} \times K_s(w, \Delta) - \frac{2}{w \times w_g} \times K_m(w, \Delta, w_g, h, h + \Delta) + \frac{1}{w_g^2} \times K_s(w_g, \Delta) \right] \qquad (2.28)$$

with

$$K_s(a,b) = \mathrm{Re} \left\{ \begin{array}{l} 4 \times [K4(a) + K4(jb)] - \\ 2[K4(a + jb) + K4(a - jb)] \end{array} \right\} + \frac{1}{3} \times \pi \times a \times b^3 \qquad (2.29)$$

$$K_m(a,b,c,d,h) =$$

$$\mathrm{Re}\left\{\begin{array}{ccc} -K4(z1+z2-z3-z4-jh)\times\left.\begin{array}{c}(a/2)\\ z1=-(a/2)\end{array}\right| \\[2ex] \left.j(b/2)\right|\ \left.(c/2)\right|\ \left.j(d/2)\right| \\ z2=-j(b/2)\left|z3=-(c/2)\right|z4=j(d/2)\end{array}\right\} \qquad (2.30)$$

$$K4(z) = \frac{z^4}{24}\times\left[\ln(z)-\frac{25}{12}\right] \qquad (2.31)$$

The distributed series resistance R is given by

$$R(f) = R'_{dc} +$$

$$\cfrac{R'_{se}(f_{se})\times\cfrac{\cfrac{\sqrt{f}}{f_{se}}+\sqrt{1+\left(\cfrac{f}{f_{se}}\right)^2}}{1+\sqrt{\cfrac{f}{f_{se}}}}-\cfrac{R'_{se}(f_{se})-R'_{dc}}{\sqrt{1+\left(\cfrac{f}{f_{.0}}\right)^2}}-R'_{dc}}{1+\cfrac{0.2}{1+\cfrac{w}{h}}\,x\ln\left(1+\cfrac{f_{se}}{f}\right)} \qquad (2.32)$$

where

$$R'_{dc} = R'_{dc,w} + R'_{dc,g} \qquad (2.33)$$

Equations (2.5) to (2.33) capture the behavior of the distributed line parameters in the quasistatic regime. To extend their applicability to include dynamic dispersion, the inductance and capacitance must be corrected as follows [8]:

$$L_{dynamic} = L \times F_L \qquad (2.34)$$

$$C_{dynamic} = C \times F_C \qquad (2.35)$$

where F_L and F_C are given by

$$F_L = F_z \times \sqrt{F_\varepsilon} \qquad (2.36)$$

and

$$F_C = \frac{\sqrt{F_\varepsilon}}{F_Z} \qquad (2.37)$$

and

$$F_\varepsilon = \frac{\varepsilon_r}{\varepsilon_{r,\text{eff},0}} - \left(\frac{\dfrac{\varepsilon_r}{\varepsilon_{r,\text{eff},0}} - 1}{1 + P} \right) \qquad (2.38)$$

where

$$P = P1 \times P2 \times \left\{ (0.1844 + P3 \times P4) \times 10 \times \frac{f \times h}{\text{GHz} \times \text{cm}} \right\}^{1.5763} \qquad (2.39)$$

with

$$P1 = 0.27488 + \left(\frac{w}{h} \right) \times \left[0.6315 + \frac{0.525}{\left(1 + 0.157 \times \dfrac{f \times h}{\text{GHz} \times \text{cm}} \right)^{20}} \right]$$
$$- 0.065683 \times \exp\left(-8.7513 \times \frac{w}{h} \right) \qquad (2.40)$$

$$P2 = 0.33622 \times \left[1 - \exp(-0.03442 \times \varepsilon_r) \right] \qquad (2.41)$$

$$P3 = 0.0363 \times \exp\left(-4.6 \times \frac{w}{h} \right)$$
$$\times \left\{ 1 - \exp\left[-\left(\frac{f \times h}{3.87 \times \text{GHz} \times \text{cm}} \right) \right]^{4.97} \right\} \qquad (2.42)$$

and

$$P4 = 1 + 2.751 \times \left\{ 1 - \exp\left[-\left(\frac{\varepsilon_r}{15.916} \right)^8 \right] \right\} \qquad (2.43)$$

$$F_z = \frac{\varepsilon_{r,\text{eff},0} \times F_\varepsilon - 1}{\varepsilon_{r,\text{eff},0} - 1} \times \frac{1}{\sqrt{F_\varepsilon}} \qquad (2.44)$$

2.2.2.2 Coplanar Waveguide Line Model

The coplanar waveguide (CPW) transmission line has been utilized extensively in the development of RF MEMS [11, 12] (see Figure 2.4). The virtues of CPW lines are well known [6, 13]. They facilitate the insertion of both series and parallel (shunt) active and passive components, as well as high circuit density. In addition, the width of its traces may be shaped to enable matching component lead widths while simultaneously keeping constant the characteristic impedance [13]. Below, we present formulas for the synthesis of CPW lines, developed by Deng [13].

Two points of departure have been developed for the synthesis of CPW lines: (1) for the width W in terms of all other parameters, and (2) for the conductor spacing S in terms of all other parameters.

CPW Trace Width Synthesis

This formulation obtains the central conductor width W given the substrate specifications ε_r and H and the desired characteristic impedance Z_0, and spacing S; it is valid as long as the following conditions are satisfied:

$$\frac{S}{H} \leq \frac{10}{3(1 + \ln \varepsilon_r)} \quad \text{and} \quad \frac{W}{H} \leq \frac{80}{3(1 + \ln \varepsilon_r)} \qquad (2.45)$$

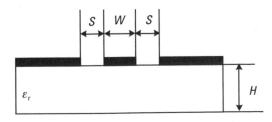

Figure 2.4 The CPW transmission line configuration.

Formulas: $W = S \times G(\varepsilon_r, H, Z_0, S)$
where

$$
G = \begin{cases} \dfrac{\left[\dfrac{1}{4}\exp\left(\dfrac{\pi}{4\sqrt{\varepsilon_{re}}}\dfrac{\eta_0}{Z_0}\right) + \exp\left(-\dfrac{\pi}{4\sqrt{\varepsilon_{re}}}\dfrac{\eta_0}{Z_0}\right) - 1\right]}{} \\ \qquad\qquad\qquad\qquad \text{for } Z_0 < \dfrac{\eta_0}{\sqrt{2(\varepsilon_r + 1)}} \\[2em] \left[\dfrac{1}{8}\exp\left(2\pi\sqrt{\varepsilon_{re}}\dfrac{Z_0}{\eta_0} - \dfrac{1}{2}\right)\right]^{-1} \quad \text{for } Z_0 \geq \dfrac{\eta_0}{\sqrt{2(\varepsilon_r + 1)}} \end{cases} \tag{2.46}
$$

where ε_{re} is the relative effective dielectric constant defined by

$$
\varepsilon_{re} = \varepsilon_{re}(\varepsilon_r, H, Z_0, S) = A \times B \tag{2.47}
$$

with

$$
A = 1 + \sqrt{2}(\varepsilon_r - 1)\sqrt{\varepsilon_r + 1}\,\frac{Z_0}{\eta_0}\frac{K(k)}{K(k')} \tag{2.48}
$$

$$
B = \operatorname{sech}\left\{\frac{\varepsilon_r^5}{4\pi(\varepsilon_r + 1)^6}\left(\frac{\eta_0}{Z_0}\right)^2 \exp\left[\begin{array}{c}\left(1 + 0.0016\varepsilon_r Z_0 \dfrac{S}{H}\right) \\ \times \ln\left(0.6 + \dfrac{S}{H}\right)\end{array}\right]\right\} \tag{2.49}
$$

$$
k = \frac{\exp\left(\dfrac{\pi(1+g)S}{2H}\right) - \exp\left(\dfrac{\pi S}{2H}\right)}{\max\left(\dfrac{\pi(1+g)S}{2H}\right) - 1}, \quad k' = \sqrt{1 - k^2} \tag{2.50}
$$

$$
g = G\Big|_{\varepsilon_{re} = \frac{\varepsilon_r + 1}{2}}, \text{ and } \eta_0 = 120\pi\Omega \tag{2.51}
$$

In (2.48), $K(k)/K(k')$ is the ratio of complete elliptic integrals of the first kind [14].

CPW Trace Spacing Synthesis

In this formulation the spacing S is obtained for given substrate parameters ε_r and H, characteristic impedance Z_0, and center conductor width W, and is valid as long as the following conditions hold:

$$\frac{W}{H} \leq \frac{80}{3(1 + \ln \varepsilon_r)} \text{ and } \frac{S}{H} \leq \frac{80}{3(1 - \ln \varepsilon_r)} \tag{2.52}$$

where ε_{re} is the relative effective dielectric constant defined by

$$\varepsilon_{re} = \varepsilon_{re}(\varepsilon_r, H, Z_0, W) = A \times B \tag{2.53}$$

Formulas: $S = W/G(\varepsilon_r, H, Z_0, W)$
  With A as given in (2.48),

$$B = 1 + \tanh\left\{ \begin{array}{c} \dfrac{\varepsilon_r^7}{4\pi^2(\varepsilon_r + 1)^8}\left(\dfrac{\eta_0}{Z_0}\right)^2 \\ \times \exp\left[\left(1 + 0.0004\varepsilon_r Z_0 \dfrac{W}{gH}\right)\ln\left(\dfrac{W}{gH}\right)\right] \end{array} \right\} \tag{2.54}$$

and

$$k = \frac{\exp\left(\dfrac{\pi(1 + 1/g)W}{2H}\right) - \exp\left(\dfrac{\pi W}{2gH}\right)}{\exp\left(\dfrac{\pi(1 + 2/g)W}{2H}\right) - 1} \tag{2.55}$$

The above CPW synthesis formulas were compared extensively by Deng [13] to both experimental data as well as numerical calculation schemes, such as the quasistatic analysis and the spectral domain full-wave analysis. Their validity and accuracy were confirmed up to 20 GHz.

2.2.3 Self-Resonance Frequency

An isolated metallic conductor trace in the presence of an ac field exhibits surface resistance and internal inductance associated with the skin effect. The complete model of the isolated trace of length l, however, must include the dc resistance and the external inductance given by

$$R_{dc} = \frac{1}{\sigma \Delta w} \tag{2.56}$$

and

$$L = \frac{1}{|I|^2} \times \int_V \mu |H|^2 \, dV \tag{2.57}$$

where L characterizes the magnetic energy storage ability of the trace in the surrounding volume V when a sinusoidal current flowing through it, of peak amplitude I produces a field of peak magnitude H.

In an actual circuit, the trace is usually part of an interconnect line and, as such, works in conjunction with another trace (e.g., the ground plane, carrying the return current). Thus, these two traces, being separated from each other by a dielectric, embody a capacitor (Figure 2.5). For a lossy dielectric, an interconnect trace of much smaller length than the operating wavelength may be represented as in Figure 2.5. Thus, we conclude that an interconnect trace can be modeled as an RLC circuit. This means that it exhibits resonance behavior at $f_0 = 1/2\pi\sqrt{(L_i + L)C_{Sub}}$. This self-resonance is intrinsic; that is, it arises because of virtually unavoidable parasitic effects. Thus, in an extreme example, if the inductance of a narrow metal trace is utilized to

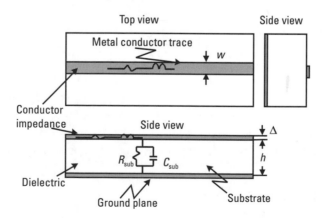

Figure 2.5 Detailed metallic conductor model.

compensate (i.e., resonate or tune out) a capacitor connected in series with it, the compensation will cease to exist beyond f_0, since in that regime the trace behaves capacitively. In a more common example, the trace might be that leading to an interdigitated monolithic microwave integrated circuits (MMIC) capacitor (Figure 2.6). At low frequencies, the lead inductance may be negligible, so it behaves as a purely ohmic wire. At sufficiently high frequencies, however, the inductive reactance of the wire may reach values that will compensate and overcome the capacitor's reactance, thus rendering the combination inductive and the capacitor inoperative. This could have unpleasant consequences, for instance, in broadband matching network applications [15].

2.2.4 Quality Factor

2.2.4.1 Definition of Quality Factor

As is well known, radio frequency and microwave circuits process information-carrying signals characterized by an incoming power level and a frequency spectrum. An unintended power loss, however, occurs when the signal passes through a passive circuit, such as an impedance matching network or a bandpass filter, due to the presence of dissipation mechanisms in its reactive elements. The quality factor Q, defined as [16]

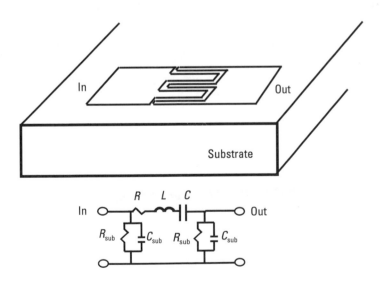

Figure 2.6   Parasitics in interdigitated planar capacitor.

$$Q = 2\pi \times \frac{\text{maximum instantaneous energy stored in the circuit}}{\text{energy dissipated per cycle}} \qquad (2.58)$$

characterizes this power loss.

In practice, the concept of Q is employed to characterize both resonant, or tuned circuits, and individual components. In the case of tuned circuits, such as a series RLC network, an easy way to obtain the Q is by the following relationship [16]:

$$Q_{\text{Tuned Circuit}} = \frac{f_0}{B} \qquad (2.59)$$

where f_0 is the resonance frequency and B is the bandwidth of the circuit. In the case of single elements, the Q is obtained by forming a resonant circuit with an ideal lossless reactance at the frequency in question [17]. Thus, for example, the Q of a capacitor with a shunt parasitic conductance G representing its losses, would be that of a tuned circuit using that capacitor and parallel conductance, with an ideal lossless inductor connected in parallel. Applying (2.58) gives

$$Q_{\text{Capacitor}} = \frac{\omega_0 C}{G} \qquad (2.60)$$

A similar analysis, applied to an inductor with series parasitic resistance R, yields an inductor Q given by

$$Q_{\text{Inductor}} = \frac{\omega_0 L}{R} \qquad (2.61)$$

A distinction is made in the literature [17] between the quality factor due to those losses intrinsic to the components, (namely, the unloaded Q) and those losses registered in the presence of coupling to an external load R_L, called loaded Q, or Q_L. The overall Q, Q_L, is expressed by

$$\frac{1}{Q_L} = \frac{1}{Q} + \frac{1}{Q_e} \qquad (2.62)$$

where $Q_e = \omega_0 L / R_L$ for a series resonant circuit, and $Q_e = R_L / \omega_0 L$ for a parallel resonant circuit. The added loss due to the external resistor, R_L, manifests itself as an insertion loss (IL) of the resonator at resonance and is given by [18]

$$IL(\text{dB}) = 20 \times \log_{10}\left(1 + \frac{Q_L}{Q}\right) \qquad (2.63)$$

2.2.4.2 On the Experimental Determination of the Quality Factor

The experimental determination of the quality factor has become the subject of great interest recently, due to its use in guiding the development and optimization of planar inductors in silicon integrated circuits [19]. In particular, it has been pointed out by O [19] that the usual formula employed,

$$Q_{\text{Meas}} = -\frac{\text{Im}(y_{11})}{\text{Re}(y_{11})} \qquad (2.64)$$

where the data for y_{11} is obtained from the measured scattering parameters of the inductors, may underestimate the actual Q, thus potentially leading to misguided inductor application and optimization. Since this formula is also equivalent to

$$Q_{\text{Meas}} = \frac{2\omega\left(\left|\overline{W}_m\right| - \left|\overline{W}_e\right|\right)}{P_{\text{diss}}} \qquad (2.65)$$

where $\left|\overline{W}_m\right|$ and $\left|\overline{W}_e\right|$ are the average stored magnetic and electrical energies in the system [1], O [19] realized that this definition is a good approximation to (2.57) only when the stored magnetic energy is much greater than the stored electrical energy. This condition is usually violated, however, in the case of silicon integrated inductors, which, possessing a large shunt capacitance to the substrate, exhibit substantial electrical energy storage. As a result, (2.64) may deviate from (2.58) by a large amount. To overcome this difficulty, O [19] proposed new methods that extract the Q by numerically adding a capacitor in parallel to measured y_{11} data of an inductor and computing the frequency stability factor and 3-dB bandwidth, (2.59), at the resonant frequency of the resulting network. Then, by computing these parameters using relationships for simple parallel RLC circuits, these parameters are converted to effective quality factors. In particular, from the formula for phase stability factor [19],

$$S_F = -\omega_0 \left.\frac{d\phi}{d\omega}\right|_{\omega=\omega_0} = -\omega_0 \left.\frac{d}{d\omega}\left[\tan^{-1}\left(\frac{\text{Im}(y_{11})}{\text{Re}(y_{11})}\right)\right]\right|_{\omega=\omega_0} \quad (2.66)$$

the Q is obtained as per (2.67):

$$Q_{\text{Effective}} = \frac{S_F}{2} \quad (2.67)$$

Equations (2.59) and (2.67) are more relevant, he points out, for circuit design, and they provide physically reasonable information throughout a wide frequency range, including the self-resonance frequency of the inductors.

2.2.4.3 Importance of Passive's Quality Factor in RF Circuits

The quality factor of passive components was introduced as a parameter to characterize the signal power loss in the component. Thus, in the context of matching networks or filters, for example, the higher the Q of the constituent passive elements, the lower the insertion loss of the circuit.

The Q, however, is also important in the context of the dc power dissipation of tuned active circuits [e.g., low-noise amplifiers (LNAs)] [20]. In this case, since the peak voltage gain is proportional to the product of transconductance and the tuned circuit's load impedance $g_m Z_0$, it is also related to the Q via $Z_0 = Q(\omega_0 L)$. Thus, from $g_m Q \omega_0 L \propto I_{DC} Q \omega_0 L/V_{GS\_Effective}$, where I_{dc} is the active device's (e.g., a FET) bias current, L the tuned circuit's load inductance, Q its quality factor, and $V_{GS\_Effective}$ the effective gate-to-source bias voltage, it becomes clear, as pointed out by Abidi, et al. [20], that there exists a trade-off between bias current, or dc power dissipation, and Q. Thus, the higher the Q, the lower the necessary bias current, and power dissipation, to achieve a given gain.

Surprisingly, perhaps, to some, there are circumstances when too high a Q in an inductor would be undesirable. This would be the case, for instance, when there is a high degree of uncertainty of the parasitics in parallel with a high-Q inductor [20]. In such a situation, the tuned circuit thus formed would be highly susceptible to frequency detuning. Adding a series resistance to the inductor would de-Q it, thus desensitizing the circuit, once tuned, to component variations. To keep the variations in gain less than 1 dB, Abidi et al. [20] determined that the loaded inductor Q must obey (2.68):

$$Q < \frac{1}{4\frac{\Delta\omega}{\omega_0}} = \frac{1}{2\sqrt{\left(\frac{\Delta L}{L}\right)^2 + \left(\frac{\Delta C}{C}\right)^2}} \qquad (2.68)$$

2.2.5 Moding (Packaging)

The proper operation of RF and microwave circuits and systems is critically dependent upon the clean environment provided by the package that houses them [2, 21, 22]. Indeed, packaging is considered an enabler for the commercialization of MEMS for at least three reasons. First, due to the sensitive nature of their moving structures, MEMS must be protected against extraneous environmental influences, such as various forms of air contamination and moisture. Second, due to their small size, it is imperative that the devices be protected to withstand handling as they are integrated with other systems. And finally, since by their very nature RF and microwave circuits and systems are susceptible to EM coupling and moding, they must be electrically isolated. Moding refers to the resonant cavity-like behavior of metal structures (Figure 2.7) enclosing high-frequency circuits, which can trap the energy being processed by these and thus contribute an effective transmission loss extrinsic to the circuit.

The problem of resonant mode suppression/damping has been studied extensively [21, 22]. Representative working solutions include that of

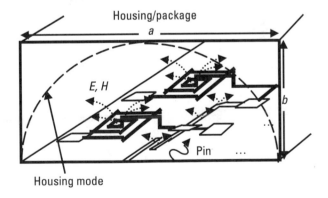

Figure 2.7 Electromagnetic environment of packaged high-frequency circuit.

Williams [21], who proposed fixing a dielectric substrate coated with a resistive film to the upper wall of the package cavity, and that of Mezzanotte et al. [22], who suggested placing damping/absorbing layers, not only on the upper wall, but also on the side walls.

2.3 Practical Aspects of RF Circuit Design

2.3.1 dc Biasing

By biasing we mean the act of interfacing a dc power source to the active devices of a circuit, such as to set them at the appropriate dc operating point, without disturbing the circuit's performance. The topic of proper biasing is important because, while ideally a voltage source should exhibit an output impedance of zero, in reality this is not the case. Thus, one must ensure that at the frequencies of interest the voltage source node does look like a short circuit.

The usual way to accomplish this is by capacitor decoupling; that is, connecting as many capacitors as necessary in parallel with the dc supply node, with values chosen so as to realize a low impedance over the frequency band of interest (Figure 2.8).

Since the capacitors themselves, by virtue of their finite quality factor and parasitic series inductance, possess extra impedance, it is necessary to calculate their value based on this net impedance:

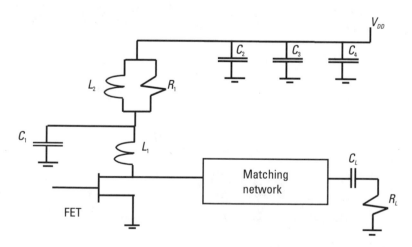

Figure 2.8 Diagram of cellular FET amplifier with emphasis on the broadband drain bias network. (*After:* [15].)

$$X_{C\_Actual} = \sqrt{(ERS)^2 + (X_L - X_C)^2} \qquad (2.69)$$

In addition to the capacitor values chosen, it is important to keep in mind the capacitor location (i.e., on-chip or off-chip). While the on-chip environment (parasitics) is usually well controlled, the off-chip environment is not (Figure 2.9). As indicated in Figure 2.9, the interface of a die to an off-chip node may include various levels of parasitic, which may give rise to multiple in-band resonances. Thus, extreme care must be exercised to accurately model the entire interface.

2.3.2 Impedance Mismatch Effects in RF MEMS

Topologically, RF/microwave systems consist of a cascade or chain connection of building blocks, with each building block performing a signal processing function (Figure 2.10). As the signal being processed propagates down the chain and reaches the input of the next building block, it is important that there be minimum reflection. Thus, as part of the system design, input and output reflection coefficients, usually expressed in terms of return loss or voltage standing wave ratio (VSWR) [the ratio between maximum (V_{max}) and minimum (V_{min}) voltage amplitudes in the interconnecting transmission line] are specified. As is well known [23], for given characteristic and load impedances, Z_0 and Z_L, respectively, the reflection coefficient, return loss and VSRW are given by

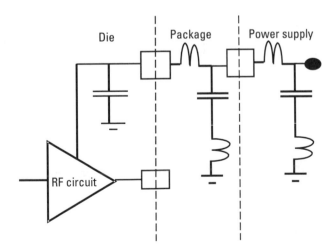

Figure 2.9 Representative on-chip-to-off-chip parasitics.

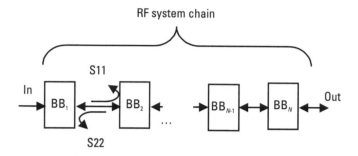

Figure 2.10 Schematic RF system chain with *N* building blocks (BB). The input (S11) and output (S22) reflection coefficients of each building block must be minimized to avoid excessively large voltage standing wave ratios (VSWRs) and, thus, voltages that might activate electrostatic MEM devices.

$$\rho = \frac{Z_0 - Z_L}{Z_0 + Z_L} \tag{2.70}$$

$$\text{Return loss (dB)} = -20 \times \log|\rho| \tag{2.71}$$

and

$$VSWR = \frac{V_{max}}{V_{min}} = \frac{1 + |\rho|}{1 - |\rho|} \tag{2.72}$$

There are a number of reasons for requiring as small input/output reflection coefficients, or equivalently as large input/output return loss, for a building block as possible. First, it is clear that very little signal power is transferred into a building block exhibiting a large input reflection coefficient (low return loss). Second, it should be clear that the bouncing back and forth of signals between building blocks might lead to destructive interference with the fresh incoming signal, thus artificially modulating the input signal to subsequent building blocks and thus causing amplitude ripples at their output. Third, from (2.72), it is clear that a large reflection coefficient implies a large ratio of maximum to minimum voltage, which, in turn, may imply a large maximum voltage amplitude. In the context of RF MEMS, inadvertently large voltage amplitudes may be undesirable if one is attempting to limit the phenomenon of hot switching. In this phenomenon, a force bias, $F = \varepsilon A V^2 / 2d^2$ [2], may be imposed on an electrostatically actuated system by virtue of the fact that the ac voltage V applied to the system is mechanically rectified to the point of inducing a bias force (voltage) greater than the pull-in force (voltage).

2.4 Problems

Problem 2.1 Skin Effect

The conductivities of aluminum, copper and gold are $\sigma_{Al} = 3.72 \times 10^7 (S/m)$, $\sigma_{Cu} = 5.8 \times 10^7 (S/m)$, and $\sigma_{Au} = 5 = 4.55 \times 10^7 (S/m)$, respectively. Calculate the skin depth at 2 GHz, 5 GHz, and 10 GHz for interconnects made out of these materials.

Solution.   With $\mu = \mu_0 = 1.26 \times 10^{-6} H/m$, and substituting in (2.1), one obtains the following:

For aluminum:

$$\delta_{2\,GHz} = \frac{1}{\sqrt{2 \times 10^9\,1/s \times \pi \times 1.26 \times 10^{-6}\,H/m \times 3.72 \times 10^7\,S/m}} \qquad (2.73)$$

$$= 1.843\,\mu m$$

$$\delta_{5\,GHz} = \frac{1}{\sqrt{5 \times 10^9\,1/s \times \pi \times 1.26 \times 10^{-6}\,H/m \times 3.72 \times 10^7\,S/m}} \qquad (2.74)$$

$$= 1.165\,\mu m$$

$$\delta_{10\,GHz} = \frac{1}{\sqrt{10 \times 10^9\,1/s \times \pi \times 1.26 \times 10^{-6}\,H/m \times 3.72 \times 10^7\,S/m}} \qquad (2.75)$$

$$= 0.824\,\mu m$$

For copper:

$$\delta_{2\,GHz} = \frac{1}{\sqrt{2 \times 10^9\,1/s \times \pi \times 1.26 \times 10^{-6}\,H/m \times 5.8 \times 10^7\,S/m}} \qquad (2.76)$$

$$= 1.476\,\mu m$$

$$\delta_{5\,GHz} = \frac{1}{\sqrt{5 \times 10^9\,1/s \times \pi \times 1.26 \times 10^{-6}\,H/m \times 5.8 \times 10^7\,S/m}}$$ (2.77)

$$= 0.933\,\mu m$$

$$\delta_{10\,GHz} = \frac{1}{\sqrt{10 \times 10^9\,1/s \times \pi \times 1.26 \times 10^{-6}\,H/m \times 5.8 \times 10^7\,S/m}}$$ (2.78)

$$= 0.66\,\mu m$$

For gold:

$$\delta_{2\,GHz} = \frac{1}{\sqrt{2 \times 10^9\,1/s \times \pi \times 1.26 \times 10^{-6}\,H/m \times 4.55 \times 10^7\,S/m}}$$ (2.79)

$$= 1.51\,\mu m$$

$$\delta_{5\,GHz} = \frac{1}{\sqrt{5 \times 10^9\,1/s \times \pi \times 1.26 \times 10^{-6}\,H/m \times 4.55 \times 10^7\,S/m}}$$ (2.80)

$$= 0.954\,\mu m$$

$$\delta_{10\,GHz} = \frac{1}{\sqrt{10 \times 10^9\,1/s \times \pi \times 1.26 \times 10^{-6}\,H/m \times 4.55 \times 10^7\,S/m}}$$ (2.81)

$$= 0.675\,\mu m$$

Problem 2.2 Self-Resonance Frequency

A 1-pF capacitor is defined as a square metallic pad on a silicon dioxide substrate. A 50Ω transmission line, providing the connection to it, is Lμm long. What is the capacitor's self-resonance frequency when (a) L = 10 μm, (b) L = 50 μm, and (c) L = 500 μm.

Solution. A transmission line of characteristic impedance Z_0, laid out on a substrate with effective dielectric constant ε_{Eff}, possesses an inductance per unit length given by

$$L_L = \frac{Z_0 \sqrt{\varepsilon_{\text{Eff}}}}{c} \qquad (2.82)$$

where $c = 3 \times 10^8$ *m/s* is the speed of light in vacuum. Using $\varepsilon_{\text{Eff}} = 3.97$ for SiO_2, we obtain an inductance per unit length of

$$L_L = \frac{50\Omega\sqrt{3.97}}{3 \times 10^8 \, m/s} = 3.32 \times 10^{-7} \, H/m = 0.332 \, pH/\mu m \qquad (2.83)$$

Thus, a 10-μm-long trace has an aggregate inductance of 3.32 pH; a 50-mm-long trace has an aggregate inductance of 16.6 pH; and a 500-mm-long trace has an aggregate inductance of 166 pH. Then, the respective self-resonance frequencies are as follows:

1. $f_{\text{Self-res}} = 87.34$ MHz;
2. $f_{\text{Self-res}} = 39.06$ MHz;
3. $f_{\text{Self-res}} = 12.35$ MHz.

Problem 2.3   dc Biasing-Decoupling

The topology shown in Figure 2.8 illustrates the elements involved in dc-biasing an active device operating at 1.9 GHz—namely, providing a low-impedance path to ground, so as to prevent RF energy from getting into the VDD supply line, and providing a high-impedance in series with the drain to prevent the in-band signal energy from being diverted to the low-impedance path and to the VDD supply line. What values should L_1, L_2, C_1, C_2, C_3, and C_4 adopt to attain broadband VDD bypass?

Solution.   With respect to Figure 2.8, a process to accomplish a broadband drain bypass scheme was detailed by Fiore [15] as follows:

1. Inductors L_1 and L_2 are utilized to provide high impedances in series with the drain, and a low-impedance path to ground is provided for them by capacitors C_1, C_2, C_3, and C_4.
2. The value of L_1 is, as a rule, picked to be 10 times higher than the transistor's drain impedance. If the drain impedance at 1.9 GHz is 9Ω, then L_1 could be chosen as $L_1 = 7.5$ nH.

3. The value of L_2 may be chosen as large as possible, consistent with self-resonance considerations.

4. C_1 is chosen as large as possible to effect a low-impedance path to ground that prevents in-band RF energy from proceeding towards the VDD supply line. A value of 10 pF gives a reactance of 9Ω at 1.9 GHz.

5. L_2, C_2, C_3, and C_4 suppress RF energy at low frequencies (i.e., at frequencies below the 1.9-GHz carrier frequency), where the amplifier is usually higher. To achieve broadband bypassing, Fiore [15] suggests staggering the values of C_2, C_3, and C_4 so that each subsequent value of impedance and inductive reactance will be low at successive frequency segments in order to enable the continuous bypassing of out-of-band frequencies below 1.9 GHz. Typical values might be $C_2 = 0.1$ uF, $C_3 = 100$ pF, $C_4 = 10$ pF, which offer reactances of 0.016Ω at 100 MHz, 3Ω at 0.5 GHz, and 11Ω at 1.5 GHz, respectively.

Problem 2.4   Quality Factor

Consider a parallel LC resonator. The equivalent parallel resistance associated with the inductor is $R_{L\_Eq} = Q_{uL}\omega L$, and the equivalent parallel resistance of the capacitor is $R_{C\_Eq} = Q_{uC}/\omega C$. (a) What is the overall equivalent parallel resistance of the LC resonator at resonance? (b) What is the overall equivalent Q? (c) If $Q_{uL} = 15$ and $Q_{uC} = 100$, what is the overall Q?

Solution.   (a) The overall equivalent parallel resistance of the LC resonator at resonance is obtained by direct substitution in the formula for combining resistors in parallel:

$$R_{Eq\_LC} = \frac{R_{Eq\_L} \times R_{Eq\_C}}{R_{Eq\_L} + R_{Eq\_C}} \tag{2.84}$$

or,

$$R_{Eq\_LC} = \frac{Q_{uL}\,\omega L \times Q_{uC}/\omega C}{Q_{uL}\,\omega L + Q_{uC}/\omega C} = \frac{Q_{uL} \times Q_{uC} \times \omega L}{Q_{uL}\,\omega L \times \omega C + Q_{uC}}$$

$$= \frac{Q_{uL} \times Q_{uC}}{Q_{uL} + Q_{uC}} = \omega L = Q_{Eq\_LC} \times \omega L \tag{2.85}$$

(b) The overall equivalent Q, from (2.83), is

$$Q_{Eq\_LC} = \frac{Q_{uL} \times Q_{uC}}{Q_{uL} + Q_{uC}} \qquad (2.86)$$

(c) For $Q_{uL} = 15$ and $Q_{uC} = 100$, we get the overall Q by substituting on (2.84). The result is $Q_{Eq\_LC} = 13.043$.

Problem 2.5   Mismatch Loss

An RF MEMS switch in its THRU state configuration exhibits a return loss of 10 dB. (a) What is its mismatch loss? (b) Would such a switch be competitive?

Solution.   (a) From (2.60), Return loss (dB) $= -20 \times \log |\rho|$, we can get the reflection coefficient as 0.316. The mismatch loss is then

$$\text{Mismatch loss (dB)} = -10 \times \log\left(1 - |\rho|^2\right) = -10$$
$$\times \log\left(1 - |0.316|^2\right) = 0.458$$

This is a measure of how much the transmitted power is attenuated due to reflection.

(b) No. RF MEMS switches that do not meet their promise of achieving ~0.1 to 2 dB insertion loss are of limited interest. To draw attention, these switches must be designed to have a mismatch loss of at most ~0.02 to 0.054 dB, which requires a return loss greater than 19 dB.

2.5   Summary

This chapter presented some critical elements for the successful implementation of RF/microwave circuits and, in particular, those exploiting RF MEMS devices. Topics included: the skin effect, which is pertinent to the aspects of dissipation and microwave energy confinement by metallic traces and shields; the revised models that are needed to properly account for the behavior of thin-substrate microstrip lines; self-resonance behavior, which is pertinent to the maximum frequency of operation of lumped passive devices; quality factor, which is important in characterizing power loss and sensitivity to parasitics in matching networks and resonators; and moding (or packaging), which is pertinent to the electromagnetic interaction between the circuit and its environment. We also dealt with two practical aspects in circuit implementation: dc biasing, which must be treated with care because it may introduce unwanted resonances that can ruin the intended frequency response; and impedance mismatches, which may cause inordinately high VSWRs and concomitantly, circuit malfunctioning by eliciting hot switching.

References

[1] Desoer, C. A., and E. S. Kuh, *Basic Circuit Theory*, New York: McGraw-Hill, 1969.

[2] De Los Santos, H. J., *Introduction to Microelectromechanical (MEM) Microwave Systems*, Norwood, MA: Artech House, 1999.

[3] Carson, R., *High-Frequency Amplifiers*, New York: John Wiley & Sons, 1975.

[4] Gonzalez, G., *Microwave Transistor Amplifiers, Analysis and Design*, Englewood Cliffs, NJ: Prentice Hall, 1984.

[5] Kraus, J. D., and K. R. Carver, *Electromagnetics*, 2nd ed., New York: McGraw-Hill, 1973.

[6] Gupta, K. C., R. Garg, and I. Bahl, *Microstrip Lines and Slot Lines*, 2nd ed., Norwood, MA: Artech House, 1996.

[7] Milanovic, V., et al., "Micromachined Microwave Transmission Lines in CMOS Technology," *IEEE Trans. Microwave Theory Tech.*, Vol. 45, pp. 630–635.

[8] Schneider, F., and W. Heinrich, "Model of Thin-Film Microstrip Line for Circuit Design," *IEEE Trans. Microwave Theory Tech.*, Vol. 49, January 2001, pp. 104–110.

[9] Bhat B., and S. K. Koul, *Stripline-like Transmission Lines for Microwave Integrated Circuits*, New Delhi, India: Wiley Eastern Limited, 1989.

[10] Djordjevic, A. R., and T. P. Sarkar, "Close-Form Formulas for Frequency-Dependent Resistance and Inductance per Unit Length of Microstrip and Strip Transmission Lines," *IEEE Trans. Microwave Theory Tech.*, Vol. 42, February 1994, pp. 241–248.

[11] Barker, N. S., and G. M. Rebeiz, "Distributed MEMS True-Time Delay Phase Shifters and Wide-Band Switches," *IEEE Trans. Microwave Theory Tech.*, Vol. 46, November 1998, pp. 1881–1890.

[12] Muldavin, J. B., and G. M. Rebeiz, "High-Isolation CPW MEMS Shunt Switches—Part 1: Modeling," *IEEE Trans. Microwave Theory Tech.*, Vol. 48, June 2000, pp. 1045–1052.

[13] Deng, T., "CAD Model for Coplanar Waveguide Synthesis," *IEEE Trans. Microwave Theory Tech.*, Vol. 44, October 1996, pp. 1733–1738.

[14] Hilberg, W., "From Approximations to Exact Relations for Characteristic Imped-ances," *IEEE Trans. Microwave Theory Tech.*, Vol. 17, May 1969, pp. 259–265.

[15] Fiore, R., "Capacitors in Broadband Applications," *Applied Microwave & Wireless*, June 2001, pp. 40–53.

[16] Krauss, H. L., C. W. Bostian, and F. H. Raab, *Solid State Radio Engineering*, New York: John Wiley & Sons, 1980.

[17] Hayward, W., *Introduction to Radio Frequency Design*, Englewood Cliffs, NJ: Prentice Hall, 1996.

[18] Ragan, G. L., (ed.), *Microwave Transmission Circuits*, Vol. 9, M.I.T. Radiation Lab Series, New York: McGraw-Hill, 1948, Ch. 10.

[19] O, K., "Estimation Methods for Quality Factors of Inductors Fabricated in Silicon Integrated Circuit Process Technologies," *IEEE J. Solid-State Circuits*, Vol. 33, August 1998, pp. 1249–1252.

[20] Abidi, A. A., G. J. Pottie, and W. J. Kaiser, "Power-Conscious Design of Wireless Cir-cuits and Systems," *Proc. IEEE*, Vol. 88, October 2000, pp. 1528–1545.

[21] Williams, D. F., "Damping of the Resonant Modes of a Rectangular Metal Package," *IEEE Trans. Microwave Theory Tech.*, Vol. 37, January 1989, pp. 253–256.

[22] Mezzanotte, P., et al., "Analysis of Packaged Microwave Integrated Circuits by FDTD," *IEEE Trans. Microwave Theory Tech.*, Vol. 42, September 1994, pp. 1796–1801.

[23] Ramo, S., J. R. Whinnery, and T. Van Duzer, *Fields and Waves on Communication Electronics*, 2nd ed., New York: John Wiley & Sons, 1984.

3

RF MEMS–Enabled Circuit Elements and Models

3.1 Introduction

The application of MEMS technology in the field of RF/microwave circuits brings to within the reach of the designer the potential to achieve unprecedented levels of performance [1–3]. Indeed, by exploiting the versatility afforded by MEMS fabrication techniques [1] (in particular, bulk and surface micromachining and LIGA) the possibility of surgically removing the perennial performance-degrading parasitics that plague monolithic RF devices becomes a reality. Furthermore, by endowing these devices with the ability to be actuated, fundamental increases in functionality become available, because now the devices become programmable, and thus, they embody not a single value, but a range of commandable values.

As suggested in the previous chapter, RF/microwave circuit design, as opposed to system design, is very physical. By this we mean that success requires an intimate awareness of both the interplay among device structure, fabrication, and physics, as well as the interactions that manifest themselves at the overall layout level. Thus, experienced RF/microwave designers have developed the ability to simultaneously think at, or visualize, multiple levels, from the fundamental/detailed to the abstract. It could even be said that in the RF/microwave regime, there is no such thing as abstract circuit design, but rather, all is ultimately concurrent device design at multiple size scales. With this in mind, this chapter presents a summary of RF MEMS devices and models, which in a logical fashion introduces their structure, circuit models, performance, and fabrication processes. Specifically, we present MEMS-based implementations of capacitors, inductors, varactors, switches, and resonators, together with their circuit design-oriented models. We conclude the chapter with a brief section on RF MEMS modeling. The RF/microwave designer will recognize that RF MEMS circuit design is, in essence, no different than conventional RF/microwave circuit design, except that in addition to the sophisticated electromagnetic phenomena that are commonly uncovered via 3-D full-wave simulation, another very important dimension—namely, that brought about by the simultaneous mechanical nature of the devices and the concomitant complexity of mechanical motion—comes into play. Thus, it will be necessary for designers to add another level to their thought processes.

3.2   RF/Microwave Substrate Properties

Invariably, planar RF/microwave devices and circuits are mechanically supported by a substrate. The nature of the substrate (i.e., whether it is conductive, semi-insulating, or insulating) plays a major role in the ultimate performance of the devices and circuits disposed on it. The quality of a substrate may, perhaps, be most easily exposed by an examination of the loss properties of a transmission line fabricated on it—in particular, a microstrip line [Figure 3.1(a)]. The loss properties of microstrip lines have been studied extensively [4–6], and they have been identified as having three sources: conduction losses in the metallic strip, dielectric losses in the substrate, and radiation losses. Metallic losses may be minimized by choosing metals with very high conductivity, and radiation losses may be minimized by eliminating the presence of sharp bends or discontinuities. Dielectric losses, however, are a direct consequence of the volume and degree of conductivity of the substrate material utilized. Indeed, from Maxwell's equation,

$$\nabla \times \vec{H} = \vec{J}_{\text{Total}} = \vec{J} + j\omega\varepsilon\vec{E} = \sigma\vec{E} + j\omega(\varepsilon' - j\varepsilon'')\vec{E}$$
$$= (\sigma + \omega\varepsilon'')\vec{E} + j\omega'\varepsilon'E = \sigma'\vec{E} + j\omega'\varepsilon'\vec{E} \qquad (3.1)$$

where $\vec{H}$ is the magnetic field of the wave propagating through the substrate, σ is the substrate conductivity, ω, ε', and ε'' are the radian frequency and real and imaginary parts of its permittivity, it is clear that power dissipation may

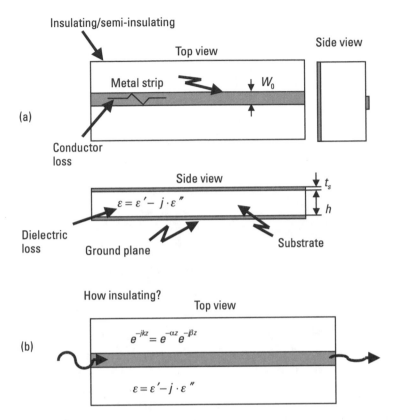

Figure 3.1 (a) Schematic of microstrip line, and (b) wave propagating down microstrip line.

arise due to the term $\sigma'\left|\vec{E}\right|^2$. At a given frequency, this, in turn, might be the result of having either a finite σ, or a finite ε'', or both. For a substrate to be considered appropriate for use in microwave circuits, therefore, it is imperative that the conduction current term in (3.1) be negligible compared to the displacement current term. This requirement is captured in the ratio $\tan\delta = \sigma'/\omega\varepsilon'$, denoted loss tangent, which, for the case of zero conductivity, becomes $\tan\delta = \varepsilon''/\varepsilon'$. Accordingly, good microwave substrates typically have $\varepsilon''/\varepsilon' \approx < 0.0001$.

From another point of view, since a wave of propagation constant $k = \alpha + j\beta$, launched with unit amplitude will, after propagating a distance z [Figure 3.1(b)], have a new reduced amplitude given by

$$e^{-jkz} = e^{-\alpha z} e^{-j\beta z} \qquad (3.2)$$

where α and β are the attenuation and phase constant, respectively, and are given by

$$\alpha = \omega \sqrt{\left(\frac{\mu \varepsilon'}{2}\right) \times \left[\sqrt{1 + \left(\frac{\varepsilon''}{\varepsilon'}\right)^2} - 1\right]} \qquad (3.3)$$

and

$$\beta = \omega \sqrt{\left(\frac{\mu \varepsilon'}{2}\right) \times \left[\sqrt{1 + \left(\frac{\varepsilon''}{\varepsilon'}\right)^2} + 1\right]} \qquad (3.4)$$

it is deduced, from an examination of (3.3), that the ratio $\varepsilon''/\varepsilon'$, the loss tangent, must be minimized, and even driven to zero, to avoid any dielectric loss. For hybrid or microwave integrated circuit (MIC) applications, where packaged discrete devices are soldered to printed interconnect lines on a substrate, this is most easily accomplished by selecting a substrate with the required loss tangent. Table 3.1 shows examples of the properties of various substrates. Unfortunately, in many occasions, as when radio-frequency integrated circuits (RFICs) must be designed in the context of established bipolar or CMOS processes, it is not possible to tailor the properties of the substrate (wafer). In such situations, the power of micromachining may be invoked.

Table 3.1
Typical Properties of Various RF/Microwave Substrates

Substrate Properties	$\varepsilon''/\varepsilon'$	Dielectric Constant (ε_r)	Resistivity (ohm-cm)
Alumina	~ 0.0002	9.9	$>10^{14}$
Aluminum nitride	~ 0.0001–0.001	8.9	$>10^{14}$
Quartz	~ 0.0004	3.8	$>10^{10}$
Intrinsic silicon	0.005	11.7	$>10^{5}$

After: [4].

3.3 Micromachined-Enhanced Elements

Micromachining, the fabrication technique to elaborate small 3-D structures in the context of planar processes, has been exploited extensively to implement high-performance passive devices [5–7]. The enhancement in RF/microwave properties usually results from the suspension of the structures, either by removal of the substrate supporting them, or from their elevation above the substrate anchoring them. Since inductors, capacitors, and transmission lines are fundamental elements in RF/microwave circuit design, many attempts at improving their properties via micromachining have been undertaken. This section presents, in a logical fashion, a representative sample of micromachining-enhanced implementations of these elements. Our goal is to develop familiarity with specific device structures, their circuit models, fabrication processes, device dimensional features, and typical performance parameters.

3.3.1 Capacitors

Capacitors are frequently employed for dc blocking and in matching networks. Two types of capacitors are normally employed in microwave circuits: (1) the interdigital capacitor for realizing values of the order of 1 pF and less, and (2) the meta-insulator-metal (MIM) capacitor for values greater than 1 pF [1]. In what follows, we examine these two structures.

3.3.1.1 Interdigitated Capacitor

The interdigitated capacitor may be modeled as in Figure 3.2, where the effective interdigital capacitance forms a series RLC circuit with the series resistance and inductance of the fingers. In addition, the conductivity and capacitance to ground introduced by the substrate contribute to diminishing its quality factor and self-resonance frequency. Clearly, since the substrate is responsible for the parasitics deteriorating the performance, it is natural to target its elimination via micromachining.

An early application of micromachining to interdigitated capacitors was performed by Chi [8], who concluded that substrate elimination should lead to improved performance and, in particular, larger quality factors due to the elimination of the substrate's capacitance to ground. In his samples, Chi did not observe any major impact on the self-resonance frequency. A subsequent investigation by Muller et al. [9], however, did find a factor-of-two improvement in self-resonance frequency (from 40 GHz to more than 100

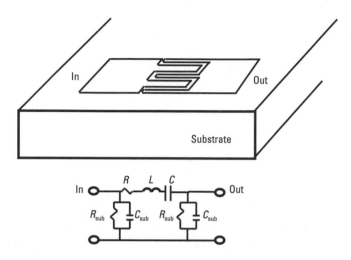

Figure 3.2 Sketch of planar interdigital capacitor and model.

GHz) upon the elimination of the substrate underneath interdigitated capacitors.

Interdigital Capacitor Fabrication

To fabricate their capacitors, Muller et al. [9] began by depositing a thin 1.5-μm dielectric membrane on a 400-μm-thick high-resistivity <100> silicon substrate (Figure 3.3). The membrane comprised a three-layer structure, including a first thermal oxide 7,000Å-thick followed by a 3000Å-thick CVD-grown Si_3N_4 layer deposited at 700°C, and finally, a 5,000Å-thick CVD layer of silicon dioxide deposited at 400°C. The membrane was suspended via backside etching, exploiting the anisotropic etching properties of KOH (potassium hydroxide).

3.3.1.2 MIM Capacitor

The MIM capacitor may be sketched and modeled as in Figure 3.4(a). Ideally, both the top and bottom plates should be isolated from the substrate; the reality, however, is that in a conventional planar process the bottom plate rests on the substrate, and consequently, it is loaded by the latter's shunt resistance and capacitance. Because of these substrate parasitics, the quality factor and self-resonance frequency may be reduced. This is particularly troublesome when the capacitor is intended for use as a series floating element, because it may not be possible to absorb the shunt capacitance into the

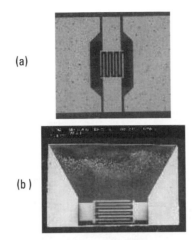

Figure 3.3 Interdigitated capacitor: (a) top view; and (b) scanning electron micrograph. The capacitor possesses eight fingers with width, spacing, and length of 25 μm, 25 μm, and 375 μm, respectively. (*Source:* [9] ©1998 IEEE. Courtesy of Dr. A. Muller, IMT—Bucharest.)

circuit. Micromachining was applied by Sun et al. [10] to improve the performance of MIM capacitors in the context of a silicon bipolar process. In this case, suspension of the MIM structure [Figure 3.4(b)] resulted in a 10-fold improvement in its quality factor—that is, from a Q of 10 to more than 100 at 2 GHz, together with a self-resonance frequency of 15.9 GHz, for a 2.6-pF capacitor.

MIM Capacitor Fabrication

To fabricate their suspended MIM capacitors, Sun et al. [10] began with a standard <100> 3.5Ω-cm silicon wafer on which was deposited a thin low-stress Si_xN_y layer for subsequent device suspension. Upon depositing a first 0.6-μm-thick metal layer, an Al_2O_3 insulating layer, and a second 1.4-μm-thick metal layer, and performing patterning, the wafers were etched from the topside using KOH. The encounter of the <100> and <111> atomic planes, as a result of the anisotropic etching, provided a natural stopping point to the process.

3.3.2 Inductors

Inductors are playing an ever-increasing role in RFICs [11, 12]. In addition to being frequently employed in passive tuning circuits or as high impedance

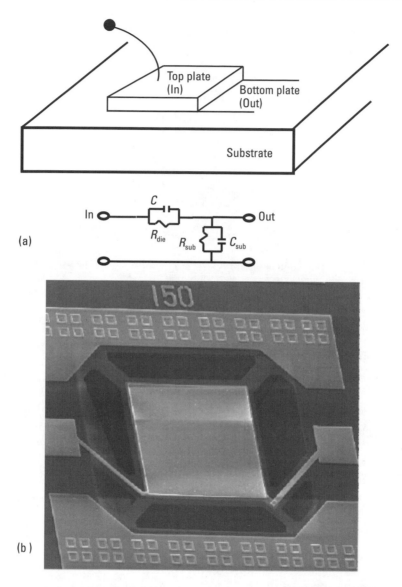

Figure 3.4 MIM capacitor: (a) schematic and circuit model; and (b) implementation. (*Source:* [10] ©1996 IEEE. Courtesy of Dr. Yanling Sun, Agere Systems.)

chokes, many novel techniques to achieve low voltage operation in advanced silicon IC processes rely on the negligible dc voltage drop across inductors [11, 12] when employed as loads or as emitter/source degenerators. When fabricated in a planar process, the trace capacitance to ground tends to lower the inductor self-resonance frequency and the substrate conductivity tends to lower its quality factor. For instance, while 25-nH spiral inductors were found to exhibit a self-resonance frequency of approximately 3 GHz when implemented on semi-insulating GaAs and insulating sapphire wafers, 10-nH spiral inductors implemented on standard silicon wafers exhibited a self-resonance frequency of only 2 GHz [13, 14].

While optimization of the spiral geometry and line width [15–17] is essential to tailor the frequency of maximum Q, this exercise only addresses minimization of the trace ohmic losses and substrate capacitance. Many attempts to use conventional processing techniques to diminish the substrate losses created by eddy currents induced by the magnetic field of the spiral have been pursued. For instance, [15] introduced blocking *p-n-p* junctions in the path of the eddy current flowing in a *p+* layer and obtained a Q improvement from 5.3 to 6 at 3.5 GHz on a 1.8-nH inductor. Yue and Wong [18], on the other hand, introduced patterned metal ground shields to block the eddy current and obtained a Q improvement from 5.08 to 6.76 at 2 GHz. While in terms of percent these improvements register as 13% and 33%, respectively, Qs greater than 10 are actually desired [11].

These limitations of planar spiral inductors have motivated the development of fabrication techniques to realize three-dimensional solenoid inductors in the context of planar processes. These inductors are expected to exhibit improved properties over their spiral counterparts because only one portion of their metal traces is susceptible to substrate capacitance. In addition, they offer design ease since an explicit relationship between their geometry and inductance value is available—namely [19],

$$L = N^2 \times \mu_{core} \times \frac{w_{core} \times h_{core}}{l_{core}} \qquad (3.5)$$

where N is the number of turns, and μ, w, h, and l are the core permeability, width, thickness, and length, respectively.

In what follows, various approaches to overcoming the limitations of silicon substrates by utilizing bulk micromachining, surface micromachining, or a combination of surface micromachining and self-assembly are addressed.

3.3.2.1　Bulk Micromachined Inductors

The pioneering work of Chang, Abidi, and Gaitan [20] first demonstrated the bulk-micromachined inductor suspended on an oxide layer and attached at four corners to the rest of the silicon wafer (Figure 3.5). The inductor, intended to have a value of 100 nH, was designed using Greenhouse's formulas [21] as a 20-turn square spiral of 4-μm-wide traces separated by 4-μm spaces, and patterned on the second-layer aluminum metal, yielding an outer side dimension of 440 μm. Calculations performed using a 3-D electromagnetic simulator came very close to measurements in predicting an increase in self-resonance frequency from 800 MHz to 3 GHz, upon substrate removal, thus validating the approach. The technique has been subsequently exploited by the same group [22], achieving a Q of 22 at 270 MHz on a 115-nH inductor.

Bulk-Micromachined Inductor Fabrication

The starting point of the suspended inductor [23] fabrication was the opening of vias on the wafer's SiO$_2$ surface passivation down to the substrate; the remaining passivation acting as a mask for the subsequent etchant. The substrate under the inductors was then removed by selective wet etching with ethylenediamine-pyrocatechol (EDP) in a procedure that lasted anywhere from 4 to 8 hours, depending on the doping level (p^- or p^+) of the substrate. While success was achieved in demonstrating the concept, several drawbacks associated with the EDP etchant were recognized. First, it has a slow, but

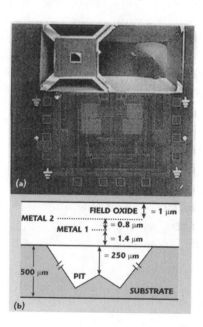

Figure 3.5   Silicon bulk-micromachined inductor. (*Source:* [19] ©1993 IEEE.)

finite aluminum etching rate; the result of which is that unprotected bond pads end up being destroyed. Second, it must be stirred, resulting in an applied pressure to the suspended membranes, which tends to shatter them upon release from the substrate. Because of these problems, a modified technique has been developed in subsequent bulk etching work [1]. In this technique, the wafers are subjected to a two-step process. In the first step, a gaseous isotropic etchant, xenon difluoride (XeF_2), is applied to create small cavities around each open area. Exposure to the gas is continued for a period of several minutes until adjacent cavities being formed beneath the surface are connected. Once this point is reached, the second step is to immerse the wafer into EDP for about 1 hour at 92°C.

3.3.2.2 Elevated-Surface Micromachined Inductors

While bulk-micromachined inductors exhibit a clearly improved performance over their conventional counterparts, a number of questions have been raised regarding (1) their mechanical ruggedness to withstand subsequent wafer processing, (2) their lack of a good RF ground, and (3) the susceptibility of their characteristics to electromagnetic coupling. The first issue is elicited upon observing the narrowness of the four beams attaching the spiral to the rest of the IC, which are indeed narrow (and mechanically weak), to minimize the device parasitic capacitance. The second issue results from the topography of the etched pit not being metallized; thus, electric field penetration into the substrate induces losses [18]. The third issue results from electromagnetic field coupling via the substrate.

To address the last two issues, Jiang et al. [24] proposed the structure shown in Figure 3.6. This structure consists of an elevated inductor suspended over a 30-μm-deep copper-lined cavity etched in the silicon bulk. The cavity depth is chosen such that the eddy currents induced in the metal shield by the magnetic field generated in the inductor are small and thus result in negligible power dissipation. This approach yielded a Q of 30 at 8 GHz on a 10.4-nH inductor, together with a self-resonance frequency of 10.1 GHz.

Elevated Inductor Fabrication

The starting substrate was a silicon wafer upon which a 650-nm-thick silicon nitride film was grown as an isolation layer. This was followed by the opening of windows, where the cavities were to be formed, by deep reactive ion etching (DRIE). The 30-μm-deep cavities thus created were then filled up via thermal oxidation. This was followed by deposition of silicon oxide to seal any remaining openings prior to the application of chemical mechanical polishing

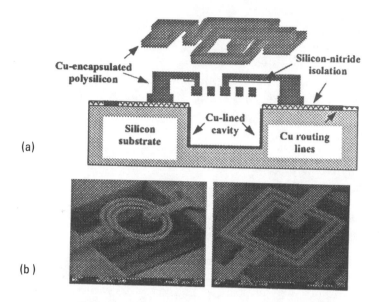

Figure 3.6 (a) Schematic of copper-encapsulated polysilicon inductor suspended over a copper-lined cavity, and (b) SEM images of circular and rectangular elevated inductor samples, each suspended over a 30-μm-deep copper-lined cavity in the silicon substrate. (*Source:* [24] ©1998 IEEE.)

(CMP) to planarize the surface. Next, the elevated spiral polysilicon structures were defined by employing conventional surface micromachining with silicon oxide as the sacrificial layer. The structures were released with hydrofluoric acid (HF).

The copper encapsulation step was carried out at a low temperature (55°C–88°C) to retain compatibility as a postprocessing step with IC fabrication. The plating was conformal and selective, with all the exposed silicon and polysilicon structures being encapsulated by the copper, while leaving areas covered by silicon oxide and silicon nitride unplated. The first step in the plating process involved a wet activation step in which the native oxide in the silicon surface was removed, with the consequent formation of an activation palladium (Pd) film on the exposed silicon and polysilicon surfaces. The effect plating the device was dipped in a solution containing cupric sulphate, formaldehyde, and ethylenediamine tetraacetic acid (EDTA) [24].

3.3.2.3 Air-Core Solenoid Inductors

While substrate removal and shielding and spiral elevation clearly improve inductor performance, the fact that there still remains a parasitic capacitance between the all-metal traces and the substrate poses the ultimate limitation on improvement. One structure that minimizes this parasitic capacitance is the solenoid. As pointed out by Yoon et al. [25], the solenoid inductor, indeed, has a number of advantages over planar inductors. First, only the bottom conductor lines are in direct contact with the substrate; thus, by making them narrow, the parasitic capacitance can be made small. Second, it is relatively easy to attain high inductance per unit area densities by simply increasing the number of turns. Third, its design, for structures exhibiting large length-to-radius ratio, is well approximated by known formulas. In this context, Yoon et al. [25] developed an approach to fabricating IC-compatible solenoids in silicon and glass substrates (Figure 3.7).

With this approach, Yoon et al. [25] decomposed the solenoid into two parts: bottom conductors and air bridges. The bottom conductor had a thickness and width of 10 and 14 μm, respectively, and the top conductor had thickness and width of 30 and 18 μm, respectively. The center-to-center height of the solenoid was 70 μm. Fabrication of the inductor on both silicon and glass substrates was demonstrated. On-silicon 2.67-nH inductors exhibited a *Q* of 16.7 at 2.4 GHz, while on-glass 2.67-nH inductors exhibited a *Q* of 24.2 at 6 GHz. Analysis of measured data revealed that the inferior properties of the on-Si inductor, relative to its on-glass counterpart, originated in the large parasitic substrate capacitance of the former. An interesting result of

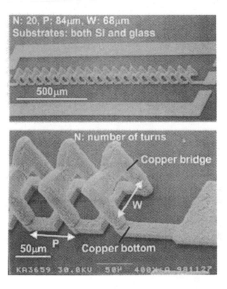

Figure 3.7 SEM photograph of 20-turn, all-copper solenoid inductor (upper: overview; lower: magnified view). (*Source:* [26] ©1998 IEEE.)

Yoon's work regarding both on-glass and on-Si implementation is that, unlike spiral, solenoid inductors exhibit a clearly linear relationship between inductance and number of turns: 0.137 nH/turn for on-glass and 0.136 nH/turn for on-Si samples. Furthermore, it was found that a narrower turn-to-turn pitch favors both higher peak Q and higher peak-Q frequency, despite a slight penalty in inductance value.

Solenoid Inductor Fabrication

Yoon et al. [26] fabricated solenoid inductors in a two-step process. First, the bottom conductors were formed via thick photoresist patterning and electroplating (Figure 3.8). Second, the air bridges were formed by a technique of multiple exposure and single development. The multiexposure consisted of performing a shallow exposure to ultraviolet (UV) light [to define the horizontal (top beam) part of the air bridges], followed by a deep exposure to create via holes for defining the anchor part. Thus exposed, the photoresist mold was completed by a standard development step. The structures were finally created by sequentially electroplating the metal to fill the via holes, overflow, and eventually form the completely connected air bridges.

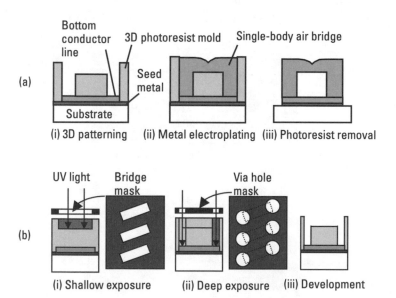

Figure 3.8 Solenoid inductor fabrication: (a) single-step air-bridge fabrication process, and (b) 3-D photolithography by multiexposure and single development. (*Source:* [25] ©1998 IEEE. Courtesy of Dr. J.-B. Yoon, KAIST.)

3.3.2.4   Embedded Solenoid Inductors

While the performance of air-core solenoid inductors is outstanding, it has been pointed out that, mechanically, they might be too delicate not only to withstand the rigors of conventional chip packaging approaches (in particular, those based on injection-molding), but to be immune to vibration (microphonics). To address these issues, Yoon et al. [27] proposed and demonstrated a scheme to produce electroplated copper solenoid inductors embedded in thick SU-8 photoresist (Figure 3.9). Their design approach was to reduce the parasitic capacitance from the coil to the substrate (by elevating the coil a distance g from the substrate) and to reduce the turn-to-turn capacitance coupling effects (by orienting the turns in parallel, spacing them an optimum distance s, and wrapping them around a core of width wc and height h). Furthermore, to make the skin effect negligible, the coil metal thickness was chosen to be 10 μm, or about 5 times the copper skin depth at 1 GHz, or 15 times that at 10 GHz. Using this approach, 2.6-nH inductors exhibited a Q of 20.5 at 4.5 GHz.

From an analysis of their measured results on inductors of varying number of turns, Yoon et al. [27] derived a new expression for characterizing the inductance of micromachined solenoid inductors:

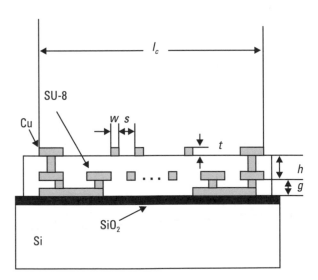

Figure 3.9 Schematic cross-sectional view of embedded solenoid inductor. The geometry on the measured device was as follows: $s = 60$ μm, $h = 40$ μm, $g = 25$ μm, $t = 10$ μm, $w = 20$ μm. (*After:* [27].)

$$L = \kappa \times N \times \mu_{core} \times \frac{w_{core} \times h_{core}}{p} \qquad (3.6)$$

where $p = l_{core}/N$ is the turn-to-turn pitch. The average inductance per turn was approximately 0.218 nH/turn.

3.3.2.5 Self-Assembled Vertical Inductors

Up to this point, all approaches to inductors that we have discussed attempt to improve performance by separating/decoupling the structure, as much as possible, from the underlying substrate. A radical approach to substrate decoupling was recently introduced by Dahlmann et al. [28], who used solder surface tension self-assembly to bring planar inductor structures perpendicular to the substrate (Figure 3.10). With this technique, separations of up to several 100 μm are possible. Indeed, measurements on 2-nH meander inductors, with an effective separation of 300 μm, indicate an improvement in Q from 4 at 1 GHz, for its planar realization, to 20 at 3 GHz, for its self-assembled implementation.

Self-Assembled Inductor Fabrication

Fabrication of the self-assembled structures [28] employs surface micromachining techniques. Essentially, the structures are patterned in a standard

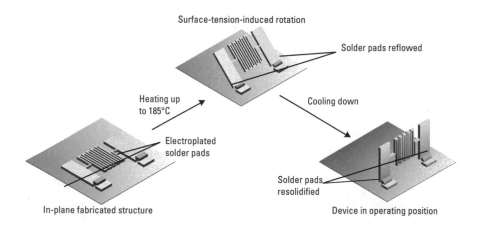

Figure 3.10 Self-assembly principle. (*Source:* [28] ©2001 Springer-Verlag. Courtesy of Dr. G. W. Dahlmann, Imperial College, London.)

planar batch process. When hinge pads are placed between a substrate anchor and a released portion of the device, subsequent heating causes the pads to melt, with the result that their surface tension force rotates the released portion out of the plane. The degree of rotation, or folding angle, of the structures is determined by the dimensions of the pads. Upon reaching the desired folding degree, cooling down the structures results in hinge resolidification and fixation of the assembly. To produce spiral inductors, the overall process requires a total of five layers of photolithography and electroplating (Figure 3.11).

3.3.3 Varactors

With the proliferation of multimode, multistandard wireless appliances, the need for high-quality varactors capable of large tunability range, at low tuning voltage spans, is a rather pressing one. Traditionally, the monolithic implementation of functions requiring tunability, such as high-performance voltage controlled oscillators (VCOs), has been precluded by the unavailability of high-quality on-chip varactors [1]. Since the varactors that are available on-chip exhibit low tuning range (on the order of 10%) and low Q, numerous efforts aimed at applying micromachining to overcome these shortcomings have been undertaken. Accordingly, based on the well-known parallel-plate capacitor equation ($C = \varepsilon A / d$), efforts have aimed at varying one of the three variables: the interplate distance d, the plate area A, or the dielectric constant ε.

3.3.3.1 Parallel Plate Varactor

An early micromachined device was the parallel-plate capacitor, demonstrated by Young and Boser [29], in which the top plate was suspended by cantilever beams (springs). Upon application of a voltage between the top and bottom plates, the electrostatic force of attraction between them overcame the stiffness of the supporting springs, thus changing the interplate distance and, in turn, changing the capacitance of the structure. The approach demonstrated a nominal 2-pF capacitor with a Q of 62 at 1 GHz, and a tuning range of 16% over 5.5V. This approach, however, is limited on two accounts. First, attaining a low tuning voltage requires soft springs, which in turn may render the structure susceptible to microphonics. Second, the pull-in effect [1] limits the theoretical tuning range to a maximum of 50% of the nominal capacitance. To improve upon this, Dec and Suyama [30] proposed and demonstrated a three-plate varactor. The structure consisted of a grounded movable metal plate sandwiched between two fixed metal plates.

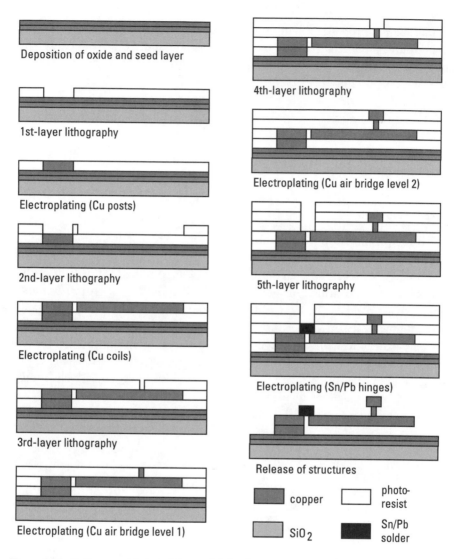

Deposition of oxide and seed layer

1st-layer lithography

Electroplating (Cu posts)

2nd-layer lithography

Electroplating (Cu coils)

3rd-layer lithography

Electroplating (Cu air bridge level 1)

4th-layer lithography

Electroplating (Cu air bridge level 2)

5th-layer lithography

Electroplating (Sn/Pb hinges)

Release of structures

copper

photo-resist

SiO_2

Sn/Pb solder

Figure 3.11 Self-assembled inductor fabrication sequence. (*Source:* [28] ©2001 Springer-Verlag. Courtesy of Dr. G. W. Dahlmann, Imperial College, London.)

As the voltage between the grounded movable plate and either of the fixed plates was varied, the movable-fixed plate capacitance changed. A prototype device, fabricated in polysilicon surface-micromachining process, exhibited a Q of 9.6 at 1 GHz for a capacitance of 4 pF, and a tuning range of 25%. A model for the parallel plate varactor was advanced by Dec and Suyama [31] (Figure 3.12).

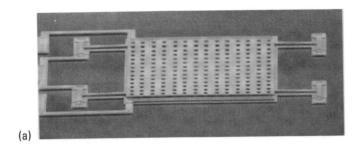

(a)

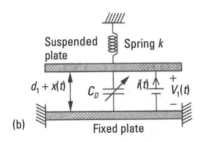

(b)

Figure 3.12 (a) Electromechanically tunable parallel-plate varactor, and (b) functional model. (*Source:* [31] ©1998 IEEE.)

In the model, the top plate is modeled as a massless structure suspended by a spring of spring constant k, and floating a distance $d + x(t)$ over a mechanically secure bottom plate. The electrostatic force of attraction elicited by applying a voltage V_{Tune} across the capacitor plates causes the suspended plate to move towards the fixed plate until the spring and electrostatic forces equilibrate. By equating these two forces, namely

$$kx = \frac{1}{2}\frac{dC_D}{dx}V_{\text{Tune}}^2 = -\frac{1}{2}\frac{\varepsilon_{\text{d}}\,AV_{\text{Tune}}^2}{(d+x)^2} \tag{3.7}$$

where ε_{d} is the dielectric constant of the interplate medium, A is the area of the capacitor plates, and d is the interplate separation, a cubic equation for $x(V)$ is obtained, whose solution is valid only for biases prior to pull-in. With $x(V)$, the capacitance is obtained from

$$C_D(V) = \frac{\varepsilon_{\text{d}}\,A}{d + x(V)} \tag{3.8}$$

3.3.3.2   Interdigitated Varactor

Addressing the issue of tunability, Yao [32] demonstrated an interdigitated varactor concept in which, instead of varying the interplate distance, what varies is the effective capacitor area A. The device was fabricated in a bulk crystalline silicon technology, which is not compatible with commercial IC processes, but exhibited excellent results: at a capacitance value of 5.19 pF, a Q of 34 at 500 MHz, a tuning range of 200% over a voltage span of 12V [32].

3.3.3.3   Movable-Dielectric Varactor

An examination of the performance-limiting mechanisms in the spring-suspended parallel-plate varactor reveals three fundamental culprits. First, its Q is limited by the series resistance of the supporting springs. Second, achieving low actuation voltage requires the use of long springs, thus imposing a conflicting trade-off with the high Q requirement. Third, the tuning range is limited by pull-in. In a novel approach, which eliminated the trade-off between Q and actuation voltage, Yoon and Nguyen [33] induced the capacitance variation by varying the effective interplate dielectric constant (Figure 3.13).

The structure consisted of fixed top and bottom capacitor plates made out of copper to minimize their total series resistance and to maximize device Q, with the movable dielectric anchored to the substrate at a point outside the two plates via spring structures. In operation [Figure 3.13(a)] the tuning dc voltage applied between the parallel plates induces charges in the dielectric, causing a pulling motion into the interplate gap, thus changing the interplate capacitance. The capacitance-voltage characteristic for the permittivity of the dielectric ε_d, which is much greater than that of air ε_a, has been shown by Yoon and Nguyen [33] to be given by

$$C = \frac{\varepsilon_a L}{t_0} + \left[\frac{\varepsilon_a \varepsilon_d}{(\varepsilon_a - \varepsilon_d)t_d + \varepsilon_d t_0} - \frac{\varepsilon_a}{t_0} \right] x =$$
$$\frac{\varepsilon_a}{t_0} \left[L + \frac{\varepsilon_a a^2}{2kt_0(1-a)^2} V_a^2 \right]$$

(3.9)

where L is the length of the plates, t_0 is the interplate distance, k is the spring constant of the dielectric suspension, t_d is the dielectric thickness, $a = t_d / t_0$, and V_a is the tuning voltage. An important aspect of the design is the fact

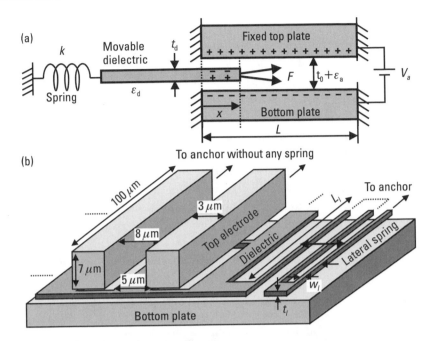

Figure 3.13 Micromachined tunable capacitor: (a) conceptual schematic; and (b) actual implementation using a lateral spring. (*Source:* [33] ©1998 IEEE. Courtesy of Dr. J.-B. Yoon, KAIST.)

that, under certain circumstances, the spring constant for motion of the dielectric perpendicular to the parallel plates may be smaller than the lateral spring constant. In that case, upon application of the bias voltage, the dielectric will displace vertically and stick to the closest plate.

The performance of a prototype sample, with nominal capacitance of 1.14 pF, characterized between 0.6 and 6 GHz, exhibited a Q of 218 at 1 GHz and zero applied voltage. Its capacitance tunability and tuning voltage span were 40% and more than 10V, respectively.

Movable-Dielectric Varactor Fabrication

There are three essential aspects of this fabrication [33]: the definition of the bottom plate, the definition of the dielectric plate and the gap between it and the plates, and the definition of the top plate. A simplified schematic of the low-temperature fabrication process (all process steps at less than 200°C for post-processing compatibility with ICs) is depicted in Figure 3.14. Beginning with a silicon wafer, a thermal SiO_2 layer is grown to provide isolation between the subsequent metal structures and the substrate. This is followed by the formation of the bottom plate, which is achieved by evaporating a 300Å/2,000Å Cr/Cu seed layer, and then electroplating copper to a thickness of 5 μm and a sheet resistance of 4.2 mΩ/sq [Figure 3.14(a)]. Next comes the deposition of a sacrificial layer between the dielectric and the bottom plate, which occurs in two steps. First, a 3,000Å-thick nickel (Ni) layer is electroplated above the bottom Cu electrode to cap it and prevent etch chamber contamination during a subsequent RIE step. Then, a 2,000Å-thick aluminum (Al) sacrificial layer is evaporated and patterned to form the vias through which the dielectric will adhere to the underlying Ni. The dielectric, PECVD nitride, is patterned via RIE to define the movable dielectric plate [Figure 3.14(b)]. Second, an additional 0.9-μm Al sacrificial layer is deposited, which, due to the topography between the fingers and the etched dielectric, ends up creating just a 0.3-μm gap between the top plate and the dielectric. Finally, as shown in Figure 3.14(c), to define the top plate, vias are

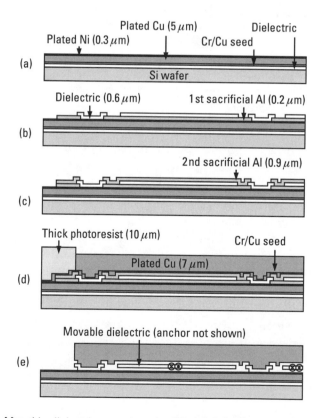

Figure 3.14 Movable-dielectric varactor simplified fabrication process. (*Source:* [33] ©1998 IEEE. Courtesy of Dr. J.-B. Yoon, KAIST.)

etched through the Al to form its anchors, followed by the evaporation of a Cr/Cu seed layer and subsequent Cu electroplating through a defining photoresist mold to a thickness of 7 mm. This thickness is determined to preclude bending of the top plates under electrostatic forces. To release the dielectric plate, the two sacrificial layers are dissolved in a $K_3Fe(CN)_6$/NaOH solution [Figure 3.14(e)]. After release, supercritical carbon dioxide drying was employed to prevent stiction. Figure 3.15 shows an SEM of the fabricated movable dielectric varactor with a lateral spring.

3.3.3.4 Digitally Controlled Parallel-Plate Varactor

The digitally controlled parallel-plate varactor represents a clever idea for extending the tunability range of parallel-plate tunable capacitors (Figure 3.16) [34]. In essence, this approach varies both the interplate distance d and the capacitor area A. Accordingly, in this structure the top plate is segmented into multiple individual plates of equal area, but suspended by springs with varying widths (and thus varying spring constants, $k_{Cantilever} = EWH^3/4L^3$). Upon actuation, the electrostatic force exerted on each plate is uniform throughout, since all the plates possess the same area and are separated the same distance from the bottom plate; but since each one is suspended by beams of different stiffness, they snap at different voltages, $V_{Pull-in} = \sqrt{8k_{Cantilever}d^3/27\varepsilon_0}$, as the applied voltage increases from zero, thus deflecting in a cascading fashion. Typical performance for a 1-pF nominal design with an interplate distance of 2 μm, exhibited a Q of 140 at 745 MHz and a

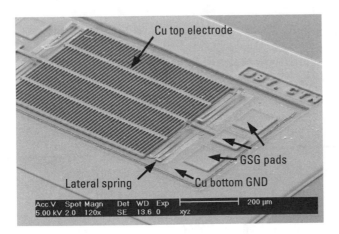

Figure 3.15 SEM of the fabricated movable dielectric varactor with a lateral spring. (*Source:* [33] ©1998 IEEE. Courtesy of Dr. J.-B. Yoon, KAIST.)

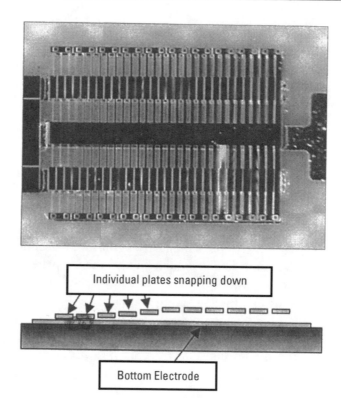

Figure 3.16 Cross-sectional view and top view picture of digitally controlled MEMS varactor. (*Source:* [34] ©2001 IEEE.)

tuning range of 400% for a voltage span of 35V. The capacitance-voltage characteristic for this device is shown in Figure 3.17.

Digitally Controlled Varactor Fabrication

The fabrication of the digitally controlled varactor combines the familiar silicon surface micromachining MUMPs process together with flip-chip bonding to an Alumina substrate via indium bums (Figure 3.18) [34].

The first step is the deposition of indium bums on gold pads defined on a prepatterned receiving substrate (e.g., Alumina) that contains the fixed bottom plate [Figure 3.18(a)]. The thickness of the bums defines the interplate distance. Then, the silicon wafer that contains the top plate with multiple snapping beams is flipped and aligned above the receiving substrate [Figure 3.18(b)]. Next, bonding is affected via thermocompression [Figure 3.18(c)]. Finally, the sacrificial oxide on the silicon wafer is dissolved by

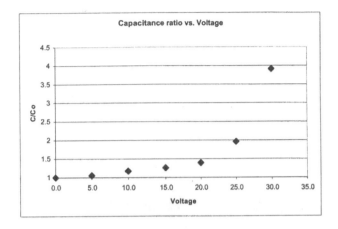

Figure 3.17 Capacitance ratio versus voltage for digitally controlled varactor. (*Source:* [34] ©2001 IEEE.)

immersion in HF to free the beams. One critical aspect of this approach is that, since the interplate distance is a function of the bum thickness, the linearity of the capacitance-voltage curve will depend on the uniformity of this thickness over the multiple plates.

3.4 MEM Switches

Switches are fundamental enablers of many RF and microwave circuits and system functions; for instance, tunable matching networks, receive/transmit switches, switching matrices, and phased array antennas [1–3]. Since MEMS technology promises to enable virtually ideal switches (in terms of power consumption, insertion loss, isolation, and linearity), extensive efforts have been aimed at their development, particularly at attaining devices exhibiting both good RF properties and low actuation voltage. The fundamentals of MEM switches, together with a plethora of implementations, are discussed in [1]. In what follows, we concentrate on the latest developments and results on the subject.

3.4.1 Shunt MEM Switch

The shunt MEM switch consists of an electrostatically actuated bridge, anchored on the ground traces of a ground-signal-ground CPW transmission line (Figure 3.19) [35].

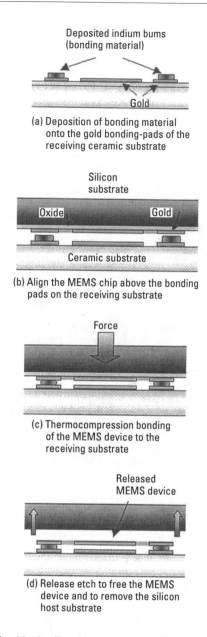

Deposited indium bums
(bonding material)

Gold

(a) Deposition of bonding material
onto the gold bonding-pads of the
receiving ceramic substrate

Silicon
substrate

Oxide   Gold

Ceramic substrate

(b) Align the MEMS chip above the bonding
pads on the receiving substrate

Force

(c) Thermocompression bonding
of the MEMS device to the
receiving substrate

Released
MEMS device

(d) Release etch to free the MEMS
device and to remove the silicon
host substrate

Figure 3.18 Steps involved in the flip-chip process used to transfer a MEMS device to a receiving substrate. (*Source:* [34] ©2001 IEEE.)

The actuation electrode, located a distance g_0 below the bridge, underneath the bridge is coated by a thin insulating layer to avoid short-circuiting

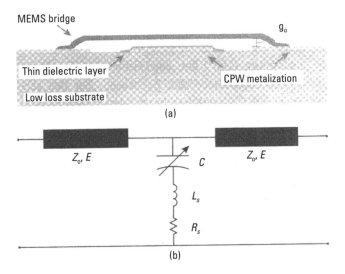

Figure 3.19 A typical shunt capacitive switch over a CPW: (a) cross section in unde-flected (ON) position, and (b) an equivalent circuit model. (*Source:* [35] ©1999 IEEE.)

upon bridge deflection. For a bottom electrode of width w, a CPW center conductor of width W, and a bridge of spring constant k, the pull-in voltage (also referred to as pull-down and actuation voltage) of the switch is given by [35]

$$V_{\text{Pull-in}} = \sqrt{\frac{8 k_{\text{Bridge}} g_0^3}{27 W w \varepsilon}} \qquad (3.10)$$

where the effective spring constant is given by [36]

$$k_{\text{Bridge}} = \frac{32 E t^3 w}{L^3} + \frac{8 \sigma (1 - v) t w}{L} \qquad (3.11)$$

where t and L are the bridge thickness and length, respectively, and E, s, and v are the Young modulus, residual tensile stress, and Poisson's ratio for the bridge material, respectively [36].

The performance level typical of these switches in the on state [37] includes an insertion loss that increases gradually with frequency from 0.1 dB below 1 GHz to about 0.3 dB at 40 GHz, a return loss of about 15 dB at

40 GHz. In the off state, the isolation ranges from nearly 0 dB at 1 GHz to about 35 dB at 40 GHz. The linearity of the switch is rather good: measurements conducted between 2 and 4 GHz failed to detect any intermodulation frequency products for signal powers ranging up to +20 dBm, yielding a third-order intercept point (IP3) of +66 dBm. Finally, the typical actuation voltage, for typical air gaps of 2 μm, insulator thickness of 0.1 μm, and dielectric constant of 7.5, lies between 30 and 50V. Herein lies the major area to which development has been addressed, namely, lowering the actuation voltage to levels compatible with mainstream IC technologies (i.e., about 5V or lower), while maintaining the RF performance substantially intact.

3.4.2 Low-Voltage Hinged MEM Switch Approaches

Examination of the equation for the pull-in of a cantilever beam, $V_{Pull-in} = \sqrt{8k_{Cantilever}d^3/27\varepsilon_0}$, reveals that the pull-in voltage may be reduced, not only by lowering the spring constant, but also by increasing the dielectric constant. Approaches that deal with both mechanisms were addressed by Park et al. [38] and Pacheco et al. [39].

3.4.2.1 Serpentine-Spring Suspended Shunt Switch

Two approaches [38, 39] have been advanced that exploit a movable plate suspended via hinges (serpentine springs), anchored on the ground traces of a CPW line, above the signal line. In one case, the movable plate was actuated by a bottom electrode, which was coated with strontium titanate oxide ($SrTiO_3$), and located underneath it. The $SrTiO_3$, evaluated individually, exhibited a relative dielectric constant between 30 and 120 (depending on the deposition temperature) and a loss tangent of less than 0.02. Among the fabricated switches, reports were given of achieving an actuation voltage as low as 8V, and measured insertion loss and isolations of 0.08 dB and 35 dB, respectively, at 10 GHz. It is not clear, however, whether all parameters were exhibited simultaneously by the same device.

In a second approach aimed at reducing the actuation voltage, Pacheco et al. [39] conceived the structure shown in Figure 3.20. This structure consists of a capacitive pad attached to actuation plates, which, in turn, are attached to folded serpentine suspensions on one end and anchored to the substrate on the other end. Upon actuation, the high capacitance of the center capacitor pad adds to the center conductor of the finite ground CPW line, causing a virtual short at high frequencies.

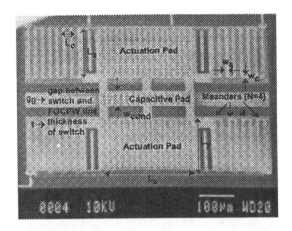

Figure 3.20 SEM of meander spring-suspended switch. Theoretical parameters: $L_s = 250$ μm, $L_c = 20\,\mu$m, $t = 2\,\mu$m, $w = 5\,\mu$m, $N = 4$, $L_x = 250\,\mu$m, $L_y = 250\,\mu$m, $w_{cond} = 60$ μm, $g_0 = 3\,\mu$m, $K_z = 0.521$ N/m, mass $= 3.23 \times 10_{-9}$ kg, $V_{Pull\text{-}in} = 1.94$V. (*Source:* [39] ©2000 IEEE.)

To lower the actuation voltage, Pacheco et al. exploited the fact that if the out-of-plane spring of a single suspension is k_z, then the effective spring constant that results when N such suspensions are connected to form N meanders is given by k_z/N, where k_z is given by

$$
k_z = \frac{Ew\left(\dfrac{t}{L_c}\right)^3}{1 + \dfrac{L_s}{L_c}\left[\left(\dfrac{L_s}{L_c}\right)^2 + 12\dfrac{1+\nu}{1+\left(\dfrac{w}{t}\right)^2}\right]}
\tag{3.12}
$$

where ν is Poisson's ratio and the rest of the variables are defined in Figure 3.20. Thus, in the case of the switch, which is supported by four such suspensions, the effective spring constant it contains is given by

$$
K_z = \frac{4k_z}{N}
\tag{3.13}
$$

Figure 3.21 shows an interesting plot of the measured structure capacitance versus voltage characteristic for various spring meanders.

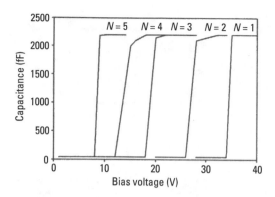

Figure 3.21 Plot of measured pull-in voltage versus capacitance with number of meanders as a parameter. (*Source:* [39] ©2000 IEEE.)

For a device with the dimensions given in the caption of Figure 3.20, an actuation voltage of 9V and insertion loss and isolation of less than 0.5 dB and about 26 dB, respectively, were obtained at 40 GHz.

Serpentine-Spring Suspended Shunt Switch Fabrication

The fabrication of the structure is accomplished with a five-mask process (Figure 3.22) [39]. Beginning with a high-resistivity 400-μm-thick silicon wafer, a 500Å/7,500Å Ti/Au layer is deposited and patterned via lift-off to define the CPW line. Next, 100Å of plasma-enhanced chemical vapor Si3N4 is deposited and patterned to define the areas underneath where insulation is required to prevent short-circuit upon deflection. To form the switch, a 3-μm-thick polyimide (DuPont PI2545) sacrificial layer is spun, soft-baked, and patterned to define the anchors. The anchors and switch are made by electroplating a 2-μm-thick Ni layer, subsequent etching of the polyimide sacrificial layer, and supercritical CO_2 drying and release of the structure.

3.4.3 Push-Pull Series Switch

As is well known [1], there is a trade-off among the RF and actuation voltage performance parameters of a MEM switch. In particular, for the series switch, while high isolation in the off state demands a large beam-to-substrate distance, a low actuation voltage demands a small beam-to-substrate distance. The push-pull [40] approach to the series switch (Figure 3.23) aims at eliminating this trade-off. The beam or lever in this structure is part of a top electrode, which, being attached to anchors via torsion springs,

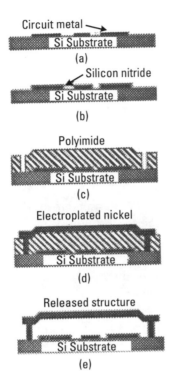

Circuit metal
Si Substrate
(a)

Silicon nitride
Si Substrate
(b)

Polyimide
Si Substrate
(c)

Electroplated nickel
Si Substrate
(d)

Released structure
Si Substrate
(e)

Figure 3.22 Serpentine-suspension switch process flow. (*Source:* [39] ©2000 IEEE.)

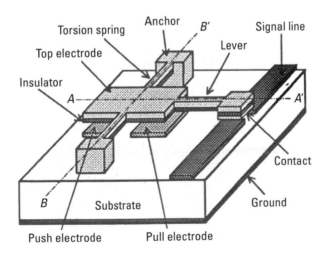

Figure 3.23 Schematic diagram of push-pull series switch. (*Source:* [40] ©2000 IEEE.)

can affect, when properly actuated, a bistable motion pivoting about the torsion spring axis. To control this pivoting motion, there are two electrodes
(the pull and the push electrodes) underneath the top electrode on either side
of the torsion axis. Thus, depending on which electrode is actuated, the lever
will move up or down. When a voltage is applied to the push electrode (Figure 3.24), the lever-to-substrate distance increases and is given by

$$h_c = \left(2 + \frac{l_{\text{lever}}}{l_{te}}\right) \times h_0 \tag{3.14}$$

Thus, according to (3.14), simultaneous low actuation voltage and
high isolation can be achieved by decreasing h_0 and increasing the lever
length l_{lever}. By choosing h_0 (the contact-to-substrate gap) such that it is
smaller than the top electrode to pull electrode gap, pull-in action during the
pull electrode actuation (turn-on process) may be avoided, as the distance h_0
may be traveled by the contact prior to the pull-in being reached under the
top-pull electrode gap. A prototype implementation of the concept (Figure
3.25) exhibited an actuation voltage of 6V and insertion loss and isolation of
about 1.5 dB and 28 dB, respectively, at 4 GHz.

Push-Pull Switch Fabrication

The fabrication process flow is shown in Figure 3.26 [40]. The beginning
wafer was GaAs, upon which the signal line and the bottom electrodes were
patterned by plating Au over a seed layer of Ti/Au [Figure 3.26(a)]. Then,
upon removal of the seed metal, the sacrificial layer and anchors are defined.
The spun photoresist (AZ5214) must be cured at 150°C in order to survive
the rest of the processing sequence. The contact is formed next by evaporating and patterning Au/Ti (0.5/0.05 μm) via a wet etch [Figure 3.26(b)].

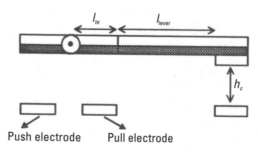

Figure 3.24 Push-pull switch configuration. (*Source:* [40] ©2000 IEEE.)

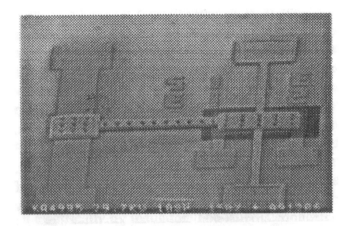

Figure 3.25 SEM micrograph of fabricated push-pull series switch. Geometrical parameters: torsion spring width, $ws = 20\ \mu$m; torsion spring length, $l_s = 300$ μm; top electrode width, $w_{te} = 100\text{–}400\ \mu$m; top electrode length, $l_{te} = 100\text{–}400\ \mu$m; bottom electrode length, $l_{be} = 0.75 l_{te}$; lever width, $w_{lever} = 50\ \mu$m; lever length, $l_{lever} = l_{te}$ to 3 l_{te}; contact-to-substrate distance, $h_0 = 1\ \mu$m. (Source: [40] ©2000 IEEE.)

Following this, the top electrode/lever movable structure, which consists of a multilayer film of SiN$_x$(200 nm)/evaporated Ti/Au(20/50 nm)/plated Au(1.1 μm) is deposited [Figure 3.26(c)]. This is followed by ground plane plating on the backside and structure release in oxygen (O$_2$) plasma dry etching to prevent stiction.

3.4.4 Folded-Beam-Springs Suspension Series Switch

The conventional series switch [1] consists of one anchored cantilever beam disposed perpendicular to a segmented transmission line. The beam usually has attached to it two isolated conductors: one that serves as the top plate of a parallel plate (driver) capacitor and one that serves as a contact (typically found attached to the underside of the beam tip). Below the beam, on the substrate surface, a third conductor, or bottom electrode, is found. Actuation of the switch is accomplished by applying a voltage of appropriate magnitude between the bottom electrode and that on the beam, which results in a downward beam deflection and ultimately causes the contact at the beam tip to bridge/closed the segmented transmission line.

One potential drawback of this structure is that, since the beam is a composite structure, the various temperature coefficients of expansion of the

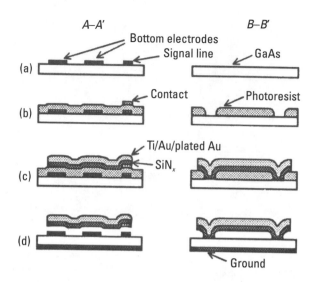

Figure 3.26   Push-pull switch fabrication process. (*Source:* [40] ©2000 IEEE.)

constituent elements may, as a result of processing conditions, result on different stress levels, which, in turn, may cause the beam to adopt a warped configuration. This makes it difficult to achieve the targeted specifications and good uniformity over the entire wafer.

A novel structure that aims at circumventing these deficiencies in the conventional cantilever beam series switch was introduced by Mihailovich et al. [41] and is shown in Figure 3.27(a).

This structure could be visualized as one in which the different parts of the switch (the actuation contacts/pads, the bridge, and the supporting cantilever beam) are decoupled. Examination of Figure 3.27(a) reveals, for instance, that the cantilever beam has been replaced by four insulating folded-beam springs. These folded beams, in turn, now support a mechanical platform, also insulating, to whose underside the isolated actuation contacts (top plates of drive capacitors) and the bridge are attached. The bridge in this case is not a smooth bar, but contains gold-based contact bums. Finally, the overall structure is anchored to the substrate by legs at the ends of the springs. The circuit model for this structure is shown in Figure 3.27(b) [42].

The performance at 40 GHz for a 500-μm-long folded-beam spring series switch is as follows [41]. In the off state, the effective capacitance is about 2 fF, for an isolation of 30 dB. In the on state, the effective resistance is about 1Ω, for an insertion loss of 0.2 dB (0.1 dB due to the contact and 0.1 dB due

(a)

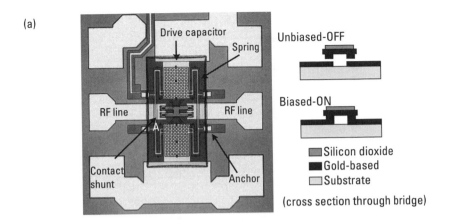

(b)

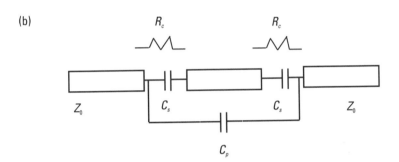

Figure 3.27 (a) Structure and operation of MEM relay. The relay consists of a metal bridge bar that is moved to make or break contact with an underlying signal line. Bridge motion is achieved by voltage biasing a mechanical actuator supporting the bridge. The device covers an area of approximately 250 × 250 μm. (*Source:* [41] ©2001 IEEE. Courtesy of Drs. R. E. Mihailovich and J. DeNatale, Rockwell Scientific.) (b) Circuit model. (*After:* [42].)

to the signal line), and the return loss is 25 dB. The actuation voltage for lowest electrical resistance is 85V, and the switching time is about 10 μs. This performance distribution of 128 measured switches is shown in Figure 3.28.

A very important parameter that characterizes the reliability of RF MEMS switches is their lifetime. Mihailovich et al. [41] conducted both hot and cold switching life tests under various signal conditions. The best results were obtained under cold conditions (no signal being switched), with typical numbers at standard ambient of 100 million cycles. Under hot-switching conditions, with the device switching a signal of 1 mA, typical numbers were in the tens of million cycles.

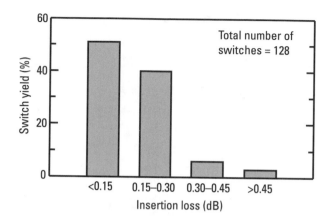

Figure 3.28 Measured RF yield at 2 GHz for an array of MEM switches. The data summa-
rizes the on-state insertion loss of 128 devices located in four fields across a
wafer. It represents an electrical yield of 90%, based upon defining a maximum insertion loss of 0.3 dB at 2 GHz with an actuation voltage of 85V, and
embodies the progress in process control methodology developed over
three device generations. (*Source:* [41] ©2001 IEEE. Courtesy of Drs. R. E.
Mihailovich and J. DeNatale, Rockwell Scientific.)

Folded-Beam-Spring Suspension Series Switch Fabrication

To fabricate the switches, Mihailovich et al. [41] employed a low temperature surface micromachining process, with all thin films deposited at temperatures below 250°C for compatibility with MMICs [1]. The substrate was
GaAs, the sacrificial layer was polyimide, and the structural layer was silicon
dioxide. The process began by defining the signal lines on the substrate via
liftoff of evaporated gold films. This was followed by spinning on the sacrificial layer, on which windows are subsequently etched to define the anchor
regions for the mechanical platform. Then, the metal bridge was patterned
on top of the polyimide by liftoff of evaporated gold. Following the metal
bridge formation, the silicon dioxide structural layer was deposited by
plasma-enhanced chemical vapor deposition (PECVD). Finally, after defining the drive metal pattern on top of the structural layer, by liftoff, and etching the PECVD film to create the beams, the entire device was released by
the isotropic dry etching of the sacrificial polyimide with O_2.

Mihailovich et al. [41] point out the degree of versatility offered by this
process as exemplified by its ability to produce devices on a variety of substrates (in particular, GaAs, epitaxial GaAs, silicon, and quartz) as long as the
substrate surface is sufficiently smooth for thin-film fabrication and can
withstand temperatures up to 250°C.

3.5   Resonators

Resonators are key elements in the realization of filters and oscillators [1], as their Q determines the insertion loss and phase noise, respectively. A number of approaches to resonators have been investigated in the context of MEMS technology: planar, volumetric or cavity resonators, micromechanical, and the film bulk surface acoustic wave (FBAR) type.

3.5.1   Transmission Line Planar Resonators

In the transmission line planar resonator, micromachining fabrication techniques are exploited to define a $\lambda/2$-long transmission line on a thin (~1.4 μm) suspended dielectric membrane [43]. Since the substrate underneath the resonator is etched, dielectric losses are eliminated. For mechanical stability, the membrane is sandwiched between lower and upper wafers that shield the structure to minimize radiation loss. Since the resonator medium is essentially air, the structure size is large, which is beneficial because the large size exhibits reduced ohmic loss [43]. A typical implementation of the resonator, with a width of 800 μm and length designed to resonate at 28.7 GHz, exhibits loaded and unloaded Qs of 190 and 460, respectively [43]. The fabrication process is similar to that discussed earlier for the interdigitated capacitor.

3.5.2   Cavity Resonators

The performance levels typical of macroscopic waveguide resonators may be approached at the microscopic planar level by exploiting micromachining techniques [2]. For example, Papapolymerou et al. [44], demonstrated an X-band micromachined cavity resonator that is suitable for integration in the context of a planar microwave process. The cavity was made by partially bulk-etching a wafer, metallizing the lower part of the cavity thus obtained, and using another wafer, with appropriate coupling slots, to cap it. In this particular demonstration, an unloaded Q of 506 for a cavity with dimensions $16 \times 32 \times 0.465$ mm was obtained. This was just 3.8% lower than the unloaded Q obtained from a rectangular cavity of identical dimensions.

One of the fundamental limitations that precludes attaining high Q is cavity volume. Recent work by Stickel, Eleftheriades, and Kremer [45] addresses this. They demonstrated a resonator with Q greater than 2,000 at 30 GHz by devising a micromachining method to construct a bigger cavity. The method involved patterning five individual wafers, etching them through from both sides in a KOH bath, and then aligning them visually using straight edge guides, prior to bonding them, both among themselves and to an unetched wafer that served as cavity bottom (Figure 3.29). The assembly was then metallized in a two-step process. First, a gold seed layer was deposited via electroless plating; second, copper was electroplated up to a 25-μm thickness.

3.5.3 Micromechanical Resonators

Micromechanical resonators offer the potential for very high Qs in the context of conventional IC processes [1]. A number of MEM resonator

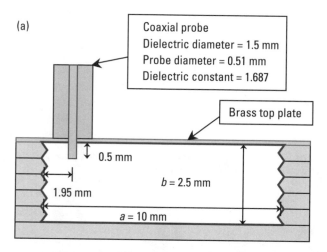

Figure 3.29 Bulk micromachined silicon cavity at 30 GHz (length $c = 5$ mm): (a) schematic; and (b) photograph. (Courtesy of Mr. M. Stickel and Prof. G. V. Eleftheriades, University of Toronto.)

structures have been investigated—namely, the clamped-clamped resonator [46], the free-free resonator [47], and the contour-mode disk resonator [48].

The clamped-clamped resonator, which has been demonstrated with frequencies around 8 MHz, has the distinction of being able to attain large stiffness-to-mass ratios, which is important for their application in communications because large stiffness enables large dynamic range and power handling [48]. Unfortunately, highly stiff clamped-clamped beams suffer from energy dissipation to the substrate via their anchors. To address this limitation, the free-free resonator, which has been demonstrated with frequencies up to 92 MHz, was proposed (Figure 3.30). In this resonator structure, the beam is suspended via four torsional beams, each of which is anchored to the

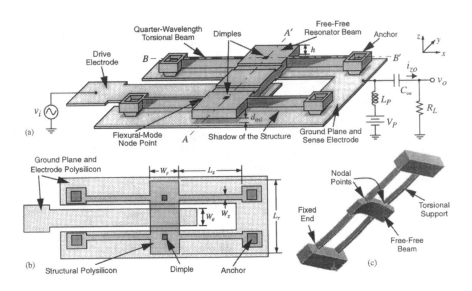

Figure 3.30 (a) Perspective view schematic of the free-free beam resonator with non-intrusive supports, explicitly indicating important features and specifying a typical bias, and off-chip output sensing configuration; (b) overhead layout view, indicating dimensions used in its design; and (c) mode shape of the resonator obtained via finite-element simulation using the ANSYS software package. (*Source:* [45] ©2000 IEEE.)

substrate via rigid contact anchors. The beams are a quarter-wavelength long such that they present a high acoustic impedance at their point of attachment to the resonator beam. This means that very little acoustic energy can propagate through the support beams to the anchors/substrate, and as a result, the Q is higher. In fact, for comparable stiffness, the Q is one order of magnitude higher than in the clamped-clamped beam resonator [48].

One key drawback of the clamped-clamped and the free-free resonators is that attaining high resonance frequencies, beyond about 92 MHz, would entail drastically scaling down their dimensions. This would result in them adopting very small sizes and exhibiting difficulty in achieving a consistent set of useful properties (e.g., Q, dynamic range, power handling) that will meet the needs of communications applications. The contour-mode disk resonator, which has been demonstrated with frequencies up to 156 MHz and a Q of 9,400, was proposed as a means of achieving high frequencies at relatively large dimensions [48]. These parameters are exhibited by a contour-mode disk resonator with a thickness of 2 μm, a radius of 17 μm, a transducer gap of 1,000Å, and an applied polarization voltage of 35V (Figure 3.31) [48].

In what follows, we present the description, operation, models, and fabrication of the clamped-clamped MEM resonator, closely following Nguyen's group paper [46]. Similar information pertaining to the free-free and contour-mode disk resonators may be found in [47, 48].

3.5.3.1   Clamped-Clamped MEM Resonator

Description and Operation

The clamped-clamped beam resonator (Figure 3.32) consists of a doubly supported cantilever beam disposed over a bottom electrode. The beam has length L_r, width W_r, and thickness h, and is made up of a material with Young's modulus E and mass density ρ. The bottom electrode has width We

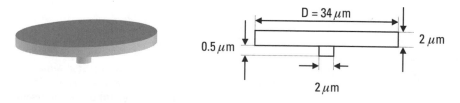

Figure 3.31   Schematic of contour-mode disk resonator. (Courtesy of Mr. Hideyuki Maekoba, Coventor, Inc.)

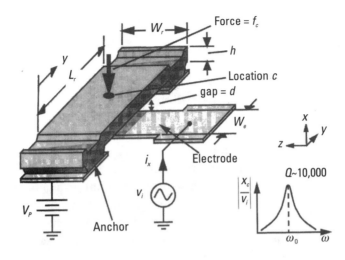

Figure 3.32 Perspective view schematic of a clamped-clamped beam MEM resonator under a typical bias and excitation configuration. Note the mechanical input force f_c. (*Source:* [45] ©2000 IEEE.)

and is separated from the beam by a gap d. In operation, a dc voltage V_P and an input ac voltage v_i, applied across the capacitor, $C = \varepsilon_0 W_r W_e / d_0$, defined by the area of overlap between the beam and the bottom electrode, induce an electrostatic force f_c on the beam, given by [46]

$$f_c = V_P \frac{\partial C}{\partial x} v_i \tag{3.15}$$

that causes it to vibrate vertically, exhibiting displacement x_c. In this expression, the derivative $\delta C / \delta x$ represents the change in electrode-to-resonator capacitance per unit displacement of the resonator and is given by

$$\frac{\partial C}{\partial x} = \frac{W_r W_e}{d_0^2} \tag{3.16}$$

where d_0 is the unbiased beam-to-electrode gap. The displacement of the beam in response to v_i induces a capacitive current given by

$$i_x = V_P \frac{\partial C}{\partial x} \frac{\partial x}{\partial t} \tag{3.17}$$

and is largest when the excitation frequency is closed to the mechanical resonance frequency of the beam, which is given by

$$f_{r,nom} = 1.03\kappa \sqrt{\frac{E}{\rho}} \times \frac{h}{L_r^2} \tag{3.18}$$

where κ is a scaling factor that models the effects of surface topography (e.g.,
the anchor step-up and its corresponding finite elasticity). In general, the
resonance frequency of the clamped-clamped beam resonator is made to
deviate from (3.18) as a result of the magnitude of the polarization voltage
V_P, which induces a dynamic [49] stiffness k_e that subtracts from the unbiased stiffness k_m. This dependence is taken into account phenomenologically
by expressing the resonance frequency by [1, 46]

$$f_0 = f_{r,nom} \sqrt{1 - \frac{k_e}{k_m}} = f_{r,nom} \sqrt{1 - g(d, V_P)} \tag{3.19}$$

where g models the effect of the electrical spring stiffness. Figure 3.33 shows
the measured resonance frequency dependence on V_P.

Physical Model

As indicated by (3.19), the spring-mass system embodied by an unbiased
clamped-clamped beam resonator exhibits a different frequency when subjected to an applied bias V_P. This implies that its equivalent mass-spring-

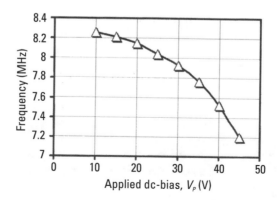

Figure 3.33 Measured frequency versus applied dc bias V_P for a parallel-plate trans-
duced clamped-clamped beam MEM resonator. (*Source:* [45] ©2000 IEEE.)

damper is bias-dependent. The corresponding bias-dependent mechanical equivalent was derived by Bannon et al. [46] in two steps: first, the unbiased parameters were derived; and second, the effect of bias was applied. With respect to Figure 3.34, the equivalent resonator mass at the location y along its length is given by the ratio of the peak kinetic energy to one-half the square of the velocity, both at y [49]:

$$m_r(y) = \frac{KE_{tot}}{(1/2)(v(y))^2} = \frac{\rho W_r h \int_0^{L_r} [X_{mode}(y')]^2 (dy')}{[X_{mode}(y)]^2} \qquad (3.20)$$

where $X_{mode}(y)$ is the shape of the fundamental mode, given by [50]

$$X_{mode}(y) = \zeta(\cos ky - \cosh ky) + (\sin ky - \sinh ky) \qquad (3.21)$$

with $k = 4.730/L_r$ and $\zeta = -1.101781$. Similarly, the equivalent spring stiffness is given by

$$k_r(y) = \omega_0^2 m_r(y) \qquad (3.22)$$

where ω_0 is the mechanical resonance frequency of the beam in radians. Finally, the equivalent damping factor is given by

$$c_r(y) = \frac{\sqrt{k_m(y)m_r(y)}}{Q_{nom}} \qquad (3.23)$$

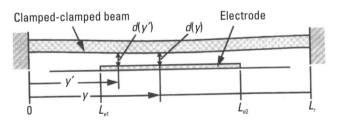

Figure 3.34 Resonator cross-section schematic for frequency-pulling and impedance analysis. (*Source:* [45] ©2000 IEEE.)

Equations (3.20) to (3.23) give the mechanical parameters for the unbiased resonator. Since the overall effect on resonance frequency of an applied dc bias is captured by (3.20), it suffices to determine the electrical spring stiffness, or the function g, in order to obtain the characterization of the biased resonator. Bannon et al. [46] defined the electrical spring stiffness k_e as resulting from the nonlinear dependence of the beam-to-electrode capacitance $C_n(x)$ on the displacement x.

Quantitatively (see Figure 3.34), the differential in electrical stiffness is given by [49]

$$dk_e(y') = V_P^2 \frac{\varepsilon_0 W_r \, dy'}{\left(d(y')\right)^3} \qquad (3.24)$$

where $d(y')$ denotes the beam-to-electrode gap resulting from the application of the dc bias V_P. By assuming that the beam deformation resulting from the polarization voltage is identical to the fundamental mode, the gap distance is approximated as [46]

$$d(y) = d_0 - \frac{1}{2} V_P^2 \varepsilon_0 W_r \int_{L_{e1}}^{L_{e2}} \frac{1}{k_m(y')\left(d(y')\right)^2} \frac{X_{\text{static}}(y)}{X_{\text{static}}(y')} dy' \quad (3.25)$$

where d_0 is the beam-to-electrode gap at $V_P = 0$. An examination of (3.25) reveals that it contains $d(y)$ on both sides, so it must be solved iteratively, beginning by assuming $d(y) = d_0$ on the right-hand side until convergence is achieved [46]. Finally, the quantity $g(d, V_P)$, which embodies the effect of the electrical stiffness, is given by

$$g(d, V_P) = \int_{L_{e1}}^{L_{e2}} \frac{dk_e(y')}{k_m} \qquad (3.26)$$

An excited beam resonator may exhibit nonlinear response due to two aspects: the nonlinear nature of its gap capacitance, and material nonlinearity due to the violation of Hooke's law. Thus, it is important to have an idea of the maximum displacement at resonance. Accordingly, at resonance the displacement at a point y along the beam is Q times the displacement under nonresonance conditions; it is given by

$$x(y) = \frac{QF_c}{k_{\text{eff}}(y)} = \frac{Q}{k_{\text{eff}}(y)} V_P \frac{\partial C}{\partial x} v_i \qquad (3.27)$$

where $k_{\text{eff}}(y)$ is the effective stiffness at location y, given by [46]

$$k_{\text{eff}}(y) = \left[\int_{L_{e1}}^{L_{e2}} \left[\frac{d_0}{d(y')} \right]^2 \frac{1}{k_r(y')} \frac{1}{W_e} \frac{X_{\text{mode}}(y)}{X_{\text{mode}}(y')} dy' \right]^{-1} \qquad (3.28)$$

To calibrate our intuition, it should be noted that a MEM resonator of 8-μm-width, 40.8-μm-length, 1.98-μm-thickness, Young's modulus of 150 GPa, mass density of 2,300 kg/m^3, gap 1,300Å, and an ac input voltage of $v_i = 3mV$ together with a dc bias voltage $v_P = 10V$, will exhibit a vibration amplitude of 49Å at the beam center. Figure 3.35 shows the transmission characteristic of a clamped-clamped resonator.

Circuit Model

Circuit design based on MEM resonators is usually carried out with the help of electromechanical analogies. Table 3.2 shows a set of such analogies. This way, the well-established CAD tools of conventional RF/microwave circuit

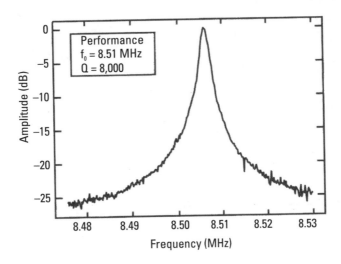

Figure 3.35 Amplitude-frequency characteristic for an 8.5-MHz polysilicon MEM resonator measured under 70 mTorr vacuum using a dc bias voltage $V_P = 10V$, a drive voltage of $v_i = 3mV$, and a transresistance amplifier with a gain of 33 $k\Omega_0$. The amplitude is v_0/v_i. (*Source:* [45] ©2000 IEEE.)

Table 3.2
Mechanical-to-Electrical Correspondence in the Current Analogy

Mechanical Variable	Electrical Variable
Damping, c	Resistance, R
Stiffness^{-1}, k^{-1}	Capacitance, C
Mass, m	Inductance, L
Force, f	Voltage, V
Velocity, v	Current, I

After: [45].

design may be exploited. Once the circuit in question meets the electrical response specifications, the analogies enable a path back to the mechanical world where the physical realization of the devices ultimately lies. A small-signal electrical equivalent circuit for the clamped-clamped MEM resonator was developed by Bannon [46] (Figure 3.36).

The resonator is modeled by a core RLC, with values given by (3.21) to (3.24), together with transformers that enable the coupling to both the electrical excitation v_i and the output force f_c.

The transformer parameters are given by

$$\eta_e = \sqrt{\int_{Le1}^{L_{e2}} \int_{Le1}^{L_{e2}} \frac{V_P^2 (\varepsilon_0 W_r)^2}{[d(y)]^2} \frac{k_{re}}{k_r(y')} \frac{X_{mode}(y)}{X_{mode}(y')} dy' dy} \qquad (3.29)$$

$$\eta_c = \sqrt{\frac{k_r\, y = 1_c}{k_{re}}} \qquad (3.30)$$

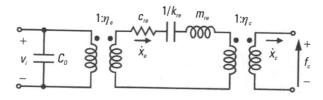

Figure 3.36 Equivalent circuit for a MEM resonator with both electrical (voltage v_i) and mechanical (force f_c) inputs and outputs, when the electrode location is assumed to be at the very center of the resonator beam. (*Source:* [45] ©2000 IEEE.)

where η_e models the transformation of input device voltage to mechanical motion, and η_c models the mechanical impedance transformation achieved by the mechanical coupling to the beam at a location a distance l_c from the anchor.

Referred to the center of the electrode port, the circuit of Figure 3.36 may be transformed into a series RLC circuit with the following values [46]:

$$L_x = \frac{m_{re}}{\eta_e^2} \qquad C_x = \frac{\eta_e^2}{k_{re}} \qquad R_x = \frac{c_{re}}{\eta_e^2} \tag{3.31}$$

where the damping constant is given by

$$c_r = \frac{\omega_0 m_r}{Q} \tag{3.32}$$

where

$$Q = Q_{nom} \sqrt{1 - g(d, V_P)} \tag{3.33}$$

Clearly, the quality factor is a function of the dynamic stiffness and, thus, of the polarization voltage V_P.

Fabrication

The fabrication of a clamped-clamped beam MEM resonator employs a polysilicon surface micromachining process (Figure 3.37) [46]. The substrate is a silicon upon which oxide and silicon nitride have been deposited; the sacrificial layer is silicon dioxide; and the structural layer is polysilicon. Wafer preparation begins by depositing isolation oxide and silicon nitride layers on a silicon wafer. This is followed by deposition of a 3,000Å-thick layer of phosphorus-doped polysilicon, following techniques employed in conventional IC processes. The polysilicon thus deposited is patterned to define the anchors, electrode, and interconnects. Then, a 1,300Å-thick sacrificial layer of LPCVD silicon dioxide is deposited and patterned. This is the cross section described by Figure 3.37(a). Next, a 2-μm-thick structural polysilicon layer is deposited by LPCVD at 585C° and doped via a POCL$_3$ gas doping step. This layer is subsequently covered by a 5,000Å LPCVD SiO$_2$ layer with a dual purpose: to serve as a dopant outdiffusion barrier during a subsequent 1-hour 1,050C° stress and dopant distribution annealing step, and to serve as

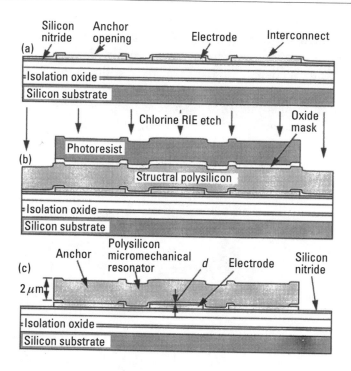

Figure 3.37 MEM resonator process sequence: (a) polysilicon electrode and interconnect layer under 1,300Å-thick sacrificial oxide; (b) required film layers and
masks needed during resonator patterning in chlorine-based RIE etch; and
(c) resulting free-standing beam following a release etch in hydrofluoric
acid. (*Source:* [45] ©2000 IEEE.)

a mask during the patterning of the polysilicon structural layer with a
chlorine-based high-density-plasma RIE. This results in the cross section
described in Figure 3.37(b). The next-to-last step is the release of the beam.
This is accomplished by immersing the wafer in a solution of hydrofluoric
acid, which preferentially etches the sacrificial oxide layer over the polysilicon
structural layer. The wafer thus released is then cleaned to remove any
remaining etch residues from the beam-to-electrode gap. This involves
immersion in a cleaning solution of H_2SO_4 and H_2O_2 (called piranha etch)
for 10 minutes, as well as a supercritical CO_2 clean.

3.5.4 Film Bulk Acoustic Wave Resonators

Currently, the maximum reported resonance frequency of MEM resonators is about 200 MHz [2]. For applications requiring resonators at higher

frequencies, particularly the 0.5 to 6 GHz range, attention is being drawn to FBARs, since they are compatible with IC fabrication processes and MEMS micromachining, and they exhibit high Qs and resonance frequencies. In addition, since acoustic waves are about five orders of magnitude shorter than electromagnetic waves, significant size reductions in the implementation of filters may be achieved [51]. For the membrane supported implementation [Figure 3.38(a)] Krishnaswamy et al. [52] demonstrated FBARs with Qs greater than 1,000 and resonance frequencies in the 1.5 to 7.5 GHz range. For the solidly mounted implementation [Figure 3.38(b)] Lakin et al. [53] demonstrated similar Qs in the 0.5 to 2.5 GHz range.

It must be stated that it is impossible to do justice to the subject of FBARs in a brief section like this. Therefore, the suggested course of action is to consult the standard texts on the subject [54, 55]. What follows is a qualitative description of FBARs in order to introduce the nomenclature and phenomenology of the device.

The FBAR is essentially an acoustic resonator cavity. Thus, the wave bouncing back and forth between walls $\lambda/2$ apart is an acoustic wave, the

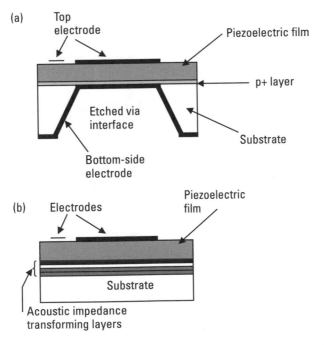

Figure 3.38 Schematic of FBAR resonators: (a) membrane supported; and (b) solidly mounted. (*After:* [51].)

walls (which also serve as electrodes) are acoustic impedance discontinuities, and the medium filling the cavity is a piezoelectric material. The structure, then, is fundamentally a capacitor whose dielectric material is piezoelectric. Excitation of the acoustic wave occurs upon application of an alternating voltage across the capacitor, which causes the piezoelectric material to cyclically expand and contract, thus eliciting energy oscillation between mechanical and electric field domains. The ratio of energy stored in the electric field U_E to that stored in acoustic field U_M is called the material's piezoelectric coupling constant and is given as

$$\frac{U_E}{U_M} = \frac{e^2}{c^E \varepsilon^S} = K^2 \qquad (3.34)$$

where e is the piezoelectric constant (which relates material strain to induced charge flux density), c^E is the material stiffness measured at a constant electric field [which relates stress to strain (Hooke's law)], and ε^S is the permittivity (measured at a constant strain). Since, in general, these constants are tensor matrices, the values to be inserted in (3.34) pertain to certain directions. The directions of interest depend upon the normal of the electrodes across which the electric field is applied and the orientation of the piezoelectric crystal. Accordingly, two modes of excitation exist: the thickness excitation (TE) mode, in which the direction of the applied electric field and the excited acoustic wave coincide; and the lateral thickness excitation (LTE) mode, in which they are perpendicular. The former is characterized by (3.34), while the latter is characterized by the electromechanical coupling constant, which is given by

$$k_t^2 = \frac{K^2}{1 + K^2} \qquad (3.35)$$

The behavior of the resonator is characterized by its impedance, which, embodying both the microwave and piezoelectric behaviors, is given by [49]

$$Z_{in} = \frac{1}{j\omega C_0} \times \left(1 - k_t^2 \times \frac{\tan(kd/2)}{kd/2}\right) \qquad (3.36)$$

where ω is the radian frequency, C_0 is the resonator parallel-plate capacitance given by

$$C_0 = \frac{\varepsilon_r \times \varepsilon_0 \times A}{d} \tag{3.37}$$

k is the acoustic wavenumber given in terms of the radian frequency and the propagation velocity v_a by

$$k = \frac{\omega}{v_a} \tag{3.38}$$

$v_a = \sqrt{c^E/\rho}$ where c^E is the stiffness and ρ the density. An examination of the input impedance expression (3.36) reveals that this becomes infinity, representing an antiresonance or parallel resonance when

$$\frac{kd}{2} = \frac{\pi}{2} \tag{3.39}$$

which occurs at a frequency ω_p, given by

$$\omega_p = \frac{\pi v_a}{d} \times N \qquad N = 1, 3, 5, \ldots \tag{3.40}$$

Similarly, the series resonance occurs when the impedance is zero; that is, when

$$1 = k_t^2 \times \frac{\tan(kd/2)}{kd/2} \tag{3.41}$$

Since (3.41) is a transcendental equation, no closed form solution exists, in general. For a small coupling constant k_t, however, an approximate expression that relates the series resonance frequency ω_s to the parallel resonance frequency has been obtained [54]:

$$\frac{\omega_p - \omega_s}{\omega_p} = \frac{4 k_t^2}{(N\pi)^2}, \qquad N = 1, 3, 5, \ldots \tag{3.42}$$

The circuit model for the intrinsic resonator is the so-called Butterworth-Van Dyke (BVD) equivalent circuit, shown in Figure 3.39.

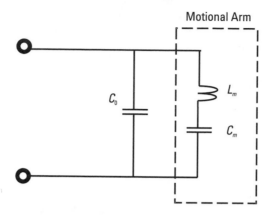

Figure 3.39 BVD equivalent circuit for ideal FBAR resonator. (*After:* [53].)

In this circuit, C_0, as previously introduced, is the parallel-plate capacitance of the resonator, and L_m and C_m determine the series and parallel resonance frequencies, respectively:

$$\omega_s = \frac{1}{\sqrt{L_m C_m}} \tag{3.43}$$

$$\omega_p = \frac{1}{\sqrt{L_m C_m}} \times \sqrt{1 + \frac{C_m}{C_0}} \tag{3.44}$$

The link between circuit and physical models is given by

$$\frac{C_m}{C_0} = \frac{8 \times k_t^2}{N^2 \pi^2} \tag{3.45}$$

and

$$L_m = \frac{\pi^3 v_a}{8\omega_s^3 \varepsilon_r \varepsilon_0 A k_t^2} \tag{3.46}$$

To account for the observed finite quality factor of the resonator, loss effects are included. These manifest as acoustic attenuation in the metallic layers and ohmic losses, and are represented by a complex propagation constant:

$$\hat{k} = \hat{k}_r + j\hat{k}_i \qquad (3.47)$$

where the real part is

$$\hat{k}_r = \frac{\omega}{v_a} \qquad (3.48)$$

and the imaginary part is

$$\hat{k}_i = \frac{\eta\omega}{2\rho v_a^2} \times \left(\frac{\omega}{v_a} \right) \qquad (3.49)$$

In (3.49), v_a is the acoustic viscosity and ρ the piezoelectric material density. In the equivalent circuit, this loss is represented by the motional resistance R_m, which is given by

$$R_m = \frac{\pi\eta\varepsilon_r\varepsilon_0}{8k_t^2 \rho A\omega v_a} \qquad (3.50)$$

When losses in the leads, R_{series}, and the parallel-plate capacitor, G_{shunt}, are taken into consideration, then the more realistic circuit in Figure 3.40 results.

From the usual expression for quality factor of a series RLC circuit $Q = \omega_s L_m / R_m$, one obtains

$$Q = \frac{v_a^2 \rho}{\omega_s \eta} \qquad (3.51)$$

An examination of the physical and circuit device models reveals that physical parameters may be inferred from circuit parameters and obtained from a curve fitting to measured data and a knowledge of the material density ρ, its dielectric constant ε_r, and its area A, as follows:

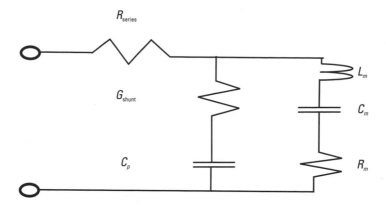

Figure 3.40 Realistic FBAR resonator equivalent circuit. Typical parameter values for a fit between 1.7 and 2.1 GHz are as follows: C_p = 1.8 pF, G_{shunt} = 1.18 S, R_{series} = 1.02 Ohms, L_m = 79.4 nH, C_m = 81 fF, R_m = 0.65 Ohm. (*After:* [55].) For AlN: ρ = 3.3g/cm^3, ε_r = 8.9.

$$k_t^2 = \frac{\pi^2}{8} \times \frac{C_m}{C_0} \tag{3.52}$$

$$v_a = \frac{L_m \times 8 \times \omega_s^3 \times \varepsilon_r \times \varepsilon_0 \times A \times k_t^2}{\pi^3} \tag{3.53}$$

$$c^E = v_a^2 \times \rho \tag{3.54}$$

$$\eta = \frac{R_m \times 8 \times k_t^2 \times \rho \times A \times \omega \times v_a}{\pi \times \varepsilon_r \times \varepsilon_0} \tag{3.55}$$

Normally, it is impossible to predict the values of process-dependent parameters such as these until the fabrication process is mature. Thus, equivalent circuit information derived from curved fitting to data from experimental devices is normally employed to elucidate these fundamental device parameters.

3.6   MEMS Modeling

The general topic of MEMS modeling is a vast and nontrivial one [57] due to the great variety of physical behavior encountered in MEMS devices. The field can be narrowed down, however, when we focus on MEMS devices for RF/microwave applications. In these cases, the great majority of devices of interest fall into one of two categories: the micromachined motionless 3-D structure, or the electrostatically actuated structure. In these contexts, then, one can posit two modeling methodologies: first principles numerical models, and reduced-order analytical (lumped-element) models. Within the former category, one finds that there exist mature, independently developed modeling tools addressing, on the one hand, the electromechanical behavior of devices (e.g., ANSYS, ABAQUS, Intellisense, CoventorWare), and on the other hand, their electromagnetic behavior (e.g., Ansoft Maxwell, Agilent HFSS). Under these circumstances (i.e., without an all-encompassing single modeling tool), RF MEMS modeling may be predicated as shown in Figure 3.41.

3.6.1 MEMS Mechanical Modeling

The mechanical modeling process begins with a statement of the desired mechanical and electrical/microwave specifications of the device. Typical mechanical specifications include actuation voltage, mechanical resonance frequency, and contact forces; while typical electrical specifications include

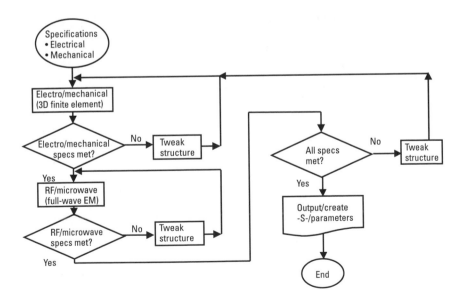

Figure 3.41 Flowchart for RF MEMS design.

scattering parameters (insertion loss, return loss, and isolation), switching time, and power dissipation–induced temperature rise.

Before detailed numerical simulation begins, approximate reduced-order analytical models are used to arrive at an approximate baseline device structure from which numerical simulations can depart.

The numerical simulation process begins with a layout of the device structure (i.e., its geometry, dimensions, and constituent materials). This description is combined with information on the fabrication process in order to emulate the effect of the process steps on the structure and to produce the 3-D solid model reflecting the process peculiarities. The solid model, thus obtained, is then meshed in preparation for the electromechanical finite-element simulation. At this point the numerical model becomes a laboratory in itself, as number runs are undertaken to explore the dependence of intended performance measures, in the context of the design space, and to arrive at the specific device design that meets the mechanical specifications. The design space typically includes geometry (i.e., structure length, width, and thickness), the dimensions of certain air gaps, the actuation area, and the effect of process-induced phenomena such as residual stress and stress gradients on performance. Further details on the mechanical design are beyond the scope of this book.

3.6.2   MEMS Electromagnetic Modeling

The electromagnetic modeling step, familiar to most RF/microwave engineers, involves the electromagnetic analysis of the structure using a 3-D full-wave solver tool. It begins by transferring the solid model developed in the mechanical simulation tool into the EM tool and proceeds with the definition of its constituent materials and boundary conditions. The analysis yields the scattering parameters and field distributions.

While the use of 3-D EM analysis is common practice in microwave design, its application to the modeling of MEMS structures is relatively recent and, indeed, meets with some challenges [58, 59] . For example, Viet-zorreck [58], who studied the modeling of the behavior of MEMS shunt capacitive switches at millimeter-wave frequencies, pointed out that these switches possess some features that make their numerical description difficult. For example, in these switches one can find, simultaneously in the direction normal to the substrate, the silicon nitride insulating the bottom electrode with a thickness of 0.1 μm, the bottom electrode with a thickness of 0.4 μm, the bridge-to-substrate distance with a thickness of 2 μm, the bridge/membrane with a thickness of 0.3 μm, the CPW ground lines with a thickness of 4 μm, the silicon dioxide insulating buffer layer with a thickness of 1 μm, and the substrate with a thickness of 545 μm. In the transverse direction, on the other hand, one finds the CPW center conductor with a width of 80 μm and a CPW ground-to-center conductor spacing of 120 μm. This coexistence of very large and very small thicknesses (e.g., 0.1 μm and 545 μm in the direction normal to the substrate) poses a limitation when it comes to discretization in the context of limited computer memory.

Another challenging aspect of the full-wave modeling of MEMS devices is that, because of their wide bandwidth, both wideband ohmic and dielectric loss behavior must be properly considered to adduce any degree of credibility to the results.

Recently, Qian et al. [60] developed a parametric model for the microwave performance of the MEMS capacitive switch in terms of a series RLC circuit (Figure 3.42).

The first step in developing the parametric model entailed performing a full wave electromagnetic simulation with the Ansoft High Frequency Structure Simulator (HFSS) on the idealized structure shown in Figure 3.42(a). In the HFSS model, the structure was enclosed in a simulation box

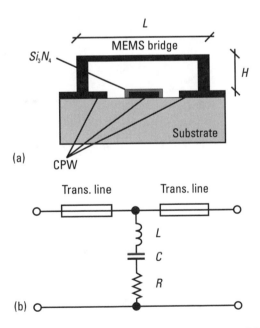

Figure 3.42 (a) Cross section of the capacitive MEMS switch over CPW line. (b) Equivalent circuit model. (*Source:* [59] ©2000 IEEE. Courtesy of Prof. F. De Flaviis and Mr. J. Qian, University of California, Irvine.)

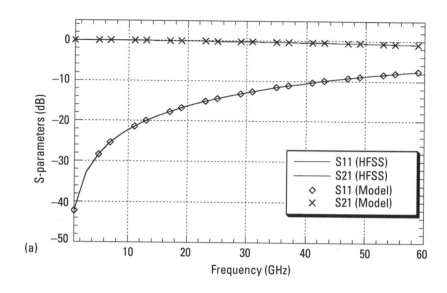

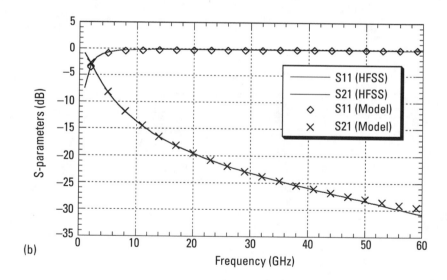

Figure 3.43 S-parameters of the EM simulated and RLC circuit model for 300-μm-long, 5-μm air-gap MEM switch: (a) switch in the off (down) state; and (b) switch in the on (up) state. (*Source:* [59] ©2000 IEEE. Courtesy of Prof. F. De Flaviis and Mr. J. Qian, University of California, Irvine.)

of size $1{,}200 \times 600 \times 600\ \mu$m on which radiation boundary conditions were imposed on all sides. The substrate was assumed to be lossless, with a relative dielectric constant of 9.8, corresponding to Alumina, and had a thickness of 600 μm. The metal structures (namely, the CPW lines and the bridge) were assumed to be perfect conductors, and the bottom electrode was assumed to be coated by a 0.1-μm-thick layer of silicon nitride with a relative dielectric constant of 7. The second step in the model development entailed running full wave simulations of the structures to generate scattering parameters in the 1 to 60 GHz frequency range. Then, via optimization, the RLC circuit parameters were extracted. The results, comparing HFSS simulations of the switch in the off and on states and the extracted model, are shown in Figure 3.43.

3.7 Summary

This chapter has presented a discussion of RF MEMS-enabled circuit elements and their models. In particular, details on a variety of capacitors, inductors, varactors, switches, and resonators, together with their physical description, fabrication process, and circuit design-oriented models, were discussed. The chapter concluded with a brief section on RF MEMS modeling, particularly as derived from the first-principles finite-element full-wave electromagnetic and mechanical computer-aided design software tools.

References

[1] De Los Santos, H. J., *Introduction to Microelectromechanical (MEM) Microwave Systems*, Norwood, MA: Artech House, 1999.

[2] Richards, R. J., and H. J. De Los Santos, "MEMS for RF/Wireless Applications: The Next Wave—Part I," *Microwave J.*, March 2001.

[3] De Los Santos, H. J, and R. J. Richards, "MEMS for RF/Wireless Applications: The Next Wave—Part II," *Microwave J.*, July 2001.

[4] http://www.mcelwee.net/html/aluminum_nitride.html.

[5] Pucel, R. H., D. J. Masse, and C. Hartwig, "Losses in Microstrip," *IEEE Trans. Microwave Theory Tech.*, Vol. 16, June 1968, pp. 342–349.

[6] Denlinger, E. J., "Losses of Microstrip," *IEEE Trans. Microwave Theory Tech.*, Vol. 28, June 1980, pp. 512–522.

[7] Goldfarb, M. E., and A. Platzker, "Losses in GaAs Microstrip," *IEEE Trans. Microwave Theory Tech.*, Vol. 28, June 1980, pp. 1957–1963.

[8] Chi, C.-Y., "Planar Microwave and Millimeter-Wave Components Using Micromachining Technologies," Ph.D. Dissertation, Radiation Laboratory, University of Michigan, 1995.

[9] Muller, A., et al., "Dielectric and Semiconductor Membranes As Support For Microwave Circuits," *Mediterranean Electrotechnical Conf., MELECON 98*, Vol. 1, 1998, pp. 387–391.

[10] Sun, Y., et al., "Suspended Membrane Inductors and Capacitors for Application in Silicon MMICs," *IEEE Microwave and Millimeter-Wave Monolithic Circuits Symposium Digest of Papers*, 1996, pp. 99–102.

[11] Abou-Allam, A., J. J. Nisbet, and M. C. Maliepaard, "Low-Voltage 1.9 GHz Front-End Receiver in 0.5m CMOS Technology," *J. Solid-State Circuits*, Vol. 36, October 2001, pp. 1434–1443.

[12] Gatta, F., et al., "A 2-dB Noise Figure 900-MHz Differential CMOS LNA," *IEEE J. Solid-State Circuits*, Vol. 36, October 2001, pp. 1444–1452.

[13] Imai, Y., M. Tokumitsu, and A. Minakawa, "Design and Performance of Low Current GaAs MMICs for L-band Front-End Applications," *IEEE Trans. Microwave Theory Tech.*, Vol. 39, February 1991, pp. 209–215

[14] Caulton, M., S. P. Knight, and D. A. Daly, "Hybrid Integrated Lumped Element Microwave Amplifiers," *IEEE Trans. Electron. Dev.*, Vol. ED-15, July 1968, pp. 459–466.

[15] Mernyei, F., et al., "Reducing the Substrate Losses of RF Integrated Inductors," *IEEE Microwave and Guided Wave Letts.*, Vol. 8, September 1998, pp. 300–301

[16] Craninckx, J., and M. S. J. Steyaert, "A 1.8 GHz Low-Phase-Noise CMOS VCO Using Optimized Hollow Spiral Inductors," *IEEE J. Solid-State Circuits*, Vol. 32, May 1997, pp. 736–744.

[17] Niknejad, A .M., and R. G. Meyer, "Analysis and Optimization of Monolithic Inductors and Transformers for RF ICs," *IEEE J. Solid-State Circuits*, Vol. 32, May 1997, pp. 375–378.

[18] Yue, C. P., and S. S. Wong, "A Study on Substrate Effects of Silicon-Based RF Passive Components," *1999 IEEE Int. Microwave Symp.*, Anaheim, CA, 1999.

[19] Kraus, J. D., and K. R. Carver, *Electromagnetics*, 2d ed., New York: McGraw-Hill, 1973.

[20] Chang, J.-Y., A. A. Abidi, and M. Gaitan, "Large Suspended Inductors on Silicon and Their Use in a 2m CMOS RF Amplifier," *IEEE Electron Device Letters*, Vol. 14, 1993, pp. 246–248.

[21] Greenhouse, H. M., "Design of Planar Rectangular Microelectronic Inductors," *IEEE Trans. Parts, Hybrids, Packaging*, Vol. PHP-10, June 1974, pp. 101–109.

[22] J. Rael, et al., "Design Methodology Used in a Single-Chip CMOS 900 Mhz Spread-Spectrum Wireless Transceiver," *Proc. Design Autimation Conf.*, June 1998, pp. 44–49.

[23] Chang, J. Y.-C., "An Integrated 900 MHz Spread-Spectrum Wireless Receiver in 1-m CMOS and a Suspended Inductor Technique," Ph.D. Dissertation, UCLA, March 1998.

[24] Jiang., H., et al., "Fabrication of High-Performance On-Chip Suspended Spiral Inductors by Micromachining and Electroless Copper Plating," *2000 IEEE IMS Digest of Papers*, Boston, MA, 2000.

[25] Yoon, J.-B, et al., "High-Performance Electroplated Solenoid-Type Integrated Inductor(SI2) for RF Applications Using Simple 3D Surface Micromachining Technology," *1998 IEEE IEDM Digest of Papers*, 1998, pp. 544–547.

[26] Yoon, J.-B., et al., "Surface Micromachined Solenoid On-Si and On-Glass Inductors for RF Applications," *IEEE Electron. Device Lett.*, Vol. 20, 1999, p. 487.

[27] Yoon, Y.-K., et al., "Embedded Solenoid Inductors for RF CMOS Power Amplifiers," *Transducers 2001 Digest of Papers*, Munich, Germany, June 2001, pp. 1114–1117.

[28] Dahlmann, G. W., et al., "MEMS High Q Microwave Inductors Using Solder Surface Tension Self-Assembly," *Transducers 2001 Digest of Papers*, Munich, Germany, June 2001, pp. 1098–1101.

[29] Young, D. J., and B. E. Boser, "A Micromachined Variable Capacitor for Monolithic Low-Noise VCOs," Solid-State Sensor and Actuator Workshop, Hilton Head, SC, June 2–6, 1996, pp. 86–89.

[30] Dec, A., and K. Suyama, "Micromachined Varactor with Wide Tuning Range," *Electron. Letts.*, Vol. 33, pp. 922–924.

[31] Dec., A., and K. Suyama, "Micromachined Electro-Mechanically Tunable Capacitors and Their Applications to RF ICs," *IEEE Trans. Microwave Theory Tech.*, Vol. 46, December 1998, pp. 2587–2596.

[32] Yao, J. J., "Topical Review: RF MEMS from a Device Perspective," *J. Micromech. Microeng*, Vol. 10, 2000, pp. R9–R38.

[33] Yoon, J.-B., and C. T.-C. Nguyen, "A High-Q Tunable Micromechanical Capacitor with Movable Dielectric for RF Applications," *1998 IEEE IEDM Digest of Papers*, pp. 489–492.

[34] Hoivik, N., et al., "Digitally Controllable Variable High-Q MEMS Capacitor for RF Applications," *2001 IEEE Int. Microwave Symp.*, Phoenix, AZ, 2001.

[35] Muldavin, J. B., and G. M. Rebeiz, "30 GHz Tuned MEMS Switches," *1999 IEEE Int. Microwave Symp.*, Anaheim, CA, 1999.

[36] Ostenberg, P., et al., "Self-consistent Simulation and Modeling of Electrostatically Deformed Diaphragms," *Proc. IEEE Microelectromechanical Systems Workshop*, January 1994, pp. 28–32.

[37] Brown, E. R., "RF-MEMS Switches for Reconfigurable Integrated Circuits," *IEEE Trans. Microwave Theory Tech.*, Vol. 46, November 1998, pp. 1868–1880.

[38] Park, J. Y., et al., "Fully Integrated Micromachined Capacitive Switches for RF Applications," *2000 IEEE Int. Microwave Symp.*, Boston, MA, 2000.

[39] Pacheco, S. P., L. P. B. Katehi, and C. T.-C. Nguyen, "Design of Low Actuation Voltage RF MEMS Switch," *2000 IEEE Int. Microwave Symp.*, Boston, MA, 2000.

[40] Hah, D., E. Yoon, and S. Hong, "A Low Voltage Actuated Micromachined Microwave Switch Using Torsion Springs and Leverage," *2000 IEEE Int. Microwave Symp.*, Boston, MA, 2000.

[41] Mihailovich, R. E., et al., "MEM Relay for Reconfigurable RF Circuits," *IEEE Microwave and Wireless Components Letts.*, Vol. 11, February 2001, pp. 53–55.

[42] Rebeiz, G. M., and J. Muldavin, "RF MEMS Switches and Switch Circuits," *IEEE Microwave Magazine*, Vol. 1, December 2001, pp.59–71.

[43] Brown, A. R., and G. M. Rebeiz, "A Ka-Band Micromachined Low-Phase Noise Oscillator," *IEEE Trans. Microwave Theory Tech.*, Vol. 47, April 1999.

[44] Papapolymerou, J., et al., "A Micromachined High-Q X-band Resonator," *IEEE Microwave and Guided Wave Letts*, Vol. 7, June 1997, pp. 168–170.

[45] Stickel, M., G. V. Eleftheriades, and P. Kremer, "A High-Q Micromachined Silicon Cavity Resonator at Ka-Band," *Electron. Lett.*, Vol. 37, No. 7, March 2001, pp. 433–435.

[46] Bannon, III, F. D., J. R. Clark, and C. T.-C. Nguyen, "High-Q HF Microelectromechanical Filters," *IEEE J. Solid-State Circuits*, Vol. 35, April 2000, pp. 512–526.

[47] Wang. K., A.-C. Wong, and C. T-C. Nguyen, "VHF Free-Free Beam High-Q Micromechanical Resonators," *ASME/IEEE J. Microelectromechanical Systems*, Vol. 9, September 2000, pp. 347–360.

[48] Clark, J. R., W.-T. Hsu, and C. T.-C. Nguyen, "High-Q VHF Micromechanical Contour-Mode Disk Resonators," *2000 IEEE Int. Electron Dev. Meeting*, pp. 493–496.

[49] Nathanson, H. C., et al., "The Resonant Gate Transistor," *IEEE Trans. Electron. Dev.*, Vol. 14, March 1967, pp. 117–133.

[50] Johnson, R. A., *Mechanical Filters in Electronics*, New York, NY: John Wiley & Sons, 1983.

[51] Horwitz, S., and C. Milton, "Applications of Film Acoustic Resonators," *1992 IEEE MTT-S Digest*, 1992.

[52] Krishnaswamy, S.V., et al., "Compact FBAR Filters Offer Low-Loss Performance," *Microwaves & RF*, September 1991, pp. 127–136.

[53] Lakin, K. M., et al., "Development of Miniature Filters for Wireless Applications," *IEEE Trans. Microwave Theory Tech.*, Vol. 43, December 1995, pp. 2933–2939.

[54] Rosenbaum, J. F., *Bulk Acoustic Wave Theory and Devices*, Norwood, MA: Artech House, 1988.

[55] Hashimoto, K., *Surface Acoustic Wave Devices in Telecommunications: Modeling and Simulation*, Berlin, Germany: Springer, 2000.

[56] Ruby, R., et al., "Ultra-Miniature High-Q Filters and Duplexers Using FBAR Technology," *2001 IEEE Int. Solid-State Circuits Conf. Digest of Papers*, San Francisco, CA, 2001.

[57] Senturia, S. D., "CAD Challenges for Microsensors, Microactuators and Microsystems," *IEEE Proc.*, Vol. 86, No. 8, 1998, pp. 1611–1126.

[58] Vietzorreck, L., "Modeling of the Millimeter-Wave Behavior of MEMS Capacitive Switches," *1999 IEEE Int. Microwave Symp.*, Anaheim, CA, 1999.

[59] Vietzorreck, L., et al., "Modeling of MEMS Capacitive Switches by TLM," *2000 IEEE Int. Microwave Symp.*, Boston, MA, 2000.

[60] Qian, J. Y., G. P. Li, and F. De Flaviis, "A Parametric Model of MEMS Capacitive Switch Operating at Microwave Frequencies," *2000 IEEE Int. Microwave Symp.*, Boston, MA, 2000.

4

Novel RF MEMS–Enabled Circuits

4.1 Introduction

For the first time in recent history, a technology is emerging that promises to enable both new paradigms in RF circuits and systems topologies and architectures as well as unprecedented levels of performance and economy. RF MEMS is widely believed to be such a technology [1–3]. There are at least two fundamental approaches for exploiting RF MEMS, namely, bottoms-up and top-down. In the bottoms-up approach, the program to be followed would involve the direct replacement of conventional circuit elements by their superior RF MEMS counterparts, with minimum or no change in circuit topology and system architecture. In the top-down approach, the designer would begin with the proverbial clean sheet of paper, unprejudiced by the limitations and constraints imposed by conventional RF technologies, and devise circuits and systems that exploit, in the fullest possible fashion, the flexibility and power of RF MEMS. In this chapter, we present a sample of novel RF MEMS–enabled functions, which are the first fruits of both approaches. The key words describing these functions are *reconfigurable* and *programmable* (i.e., these elements, endowed by RF MEMS with the ability to be actuated, embody not just a single value, but a range of commandable values). These properties, in turn, will be exhibited by the circuits and systems of which they are a part. Since most of these visionary ideas and techniques have, understandably, anticipated RF MEMS technology maturation, the presentation will be largely of a descriptive and qualitative nature.

Nevertheless, it is expected that the ideas we have chosen to present will form the pillars upon which the paradigm of ubiquitous connectivity will be built. In order to emphasize the various ideas, the chapter is organized into three main sections dealing with circuit elements (devices), circuits, and systems.

4.2 Reconfigurable Circuit Elements

At the fundamental level, passive RF/microwave circuits and systems consist of switches, capacitors, inductors, transmission lines, and resonators. These devices/circuit elements were the first on which the impact and implications of RF MEMS were studied. The MEM switch, in particular, due to its noninvasive properties (i.e., virtually ideal insertion loss and isolation [1]), may be considered the most fundamental building block, or single device, embodying reconfigurability. Thus, all other reconfigurable functions will simply embody an appropriately organized set of interconnected switches. In what follows, beginning with a function employing a single switch, we expose some of the ideas elicited by the study of MEMS-based reconfigurability.

4.2.1 The Resonant MEMS Switch

One of the key characteristics of RF MEMS switches is their ability to maintain good isolation over wide bandwidths (e.g., 35 dB at 40 GHz for CPW membrane MEMS switches [4]). These capacitive switches, however, exhibit poor isolation at low frequencies because in that regime the impedance of the membrane-to-center-conductor capacitance is not low enough. To overcome this limitation in this type of switch, Peroulis et al. [5] proposed the resonant MEMS switch, as shown in Figure 4.1.

The resonant MEMS switch [5] emerged from the observation that the parasitic inductance, in series with the membrane/plate capacitance, which derives from the supporting structure itself, narrows the bandwidth while enhancing the isolation of the switch. Peroulis et al. [5] exploited this realization by purposefully introducing inductive connecting beams to link the center capacitor plate to the rest of the structure (Figure 4.1). By designing the connecting inductance so as to resonate the plate capacitance at a certain frequency, the above-mentioned impedance is minimized and, consequently, the isolation is increased at that frequency. The frequency of highest isolation becomes

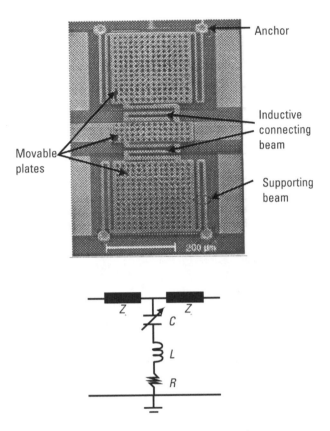

Figure 4.1 Resonant MEMS switch. (*Source:* [5] ©2001 IEEE.)

$$f_R = \frac{1}{2\pi \sqrt{C_D L}} \qquad (4.1)$$

where C_D is the plate capacitance in the down state, given by

$$C_D = \varepsilon_0 \varepsilon_r \frac{A}{t_d + t_{roughness}}; \qquad \text{down state} \qquad (4.2)$$

where A is the area of overlap of the bridge with the center conductor of the CPW line, t_d is the thickness of the dielectric protecting the bottom electrode, ε_r is its relative dielectric constant. L is the parasitic inductance, which can be determined either from measurements or from a full-wave electro-magnetic simulation of the structure and has typical values of 0.2 to 6 pH [6]. In demonstrating the concept, Peroulis et al. [5] were able to show an isolation in the down state as high as approximately 24 dB at 13.2 GHz, with inductive connecting beams of 50 pF, in a switch 3-μm air-gap structure, over a 40/60/40-μm CPW line, with a 1,000Å-thick Si_xN_y insulating layer.

4.2.2   Capacitors

Variable capacitors, or varactors, were discussed at length in Chapter 3. In this chapter, we revisit the subject of varactors, but in the context of imple-mentations that emphasize their application in circuits other than voltage-controlled oscillators (e.g., filters).

4.2.2.1   The Binary Capacitor

The binary capacitor function, a capacitance that is made to change between two values, is embodied by the shunt capacitive MEM switch. Indeed, as is well known [7], the operation of the shunt capacitive MEM switch is predi-cated upon the fact that, when it is in the up state, the capacitance from the bridge to the center conductor of the CPW line is very small; whereas when it is in the down state, that capacitance is very large. Thus, when examined, not from the insertion loss/isolation perspective, but from the perspective of the equivalent RF behavior of the structure, it may be readily characterized as a binary capacitor and exploited for tuning/reconfigurability purposes. Recently, Peroulis et al. [8] employed the serpentine-spring-supported low-voltage capacitive MEM switch [9] to demonstrate just such a binary capaci-tor (Figure 4.2). The structure's intrinsic capacitance is given by

$$C_p = \begin{cases} \varepsilon_0 \dfrac{A}{d + \dfrac{t_d}{\varepsilon_r}} + C_{\text{fringing}} ; & \text{up state} \\[2em] \varepsilon_0 \varepsilon_r \dfrac{A}{t_d + t_{\text{roughness}}} ; & \text{down state} \end{cases} \tag{4.3}$$

where A is the area of overlap of the bridge with the center conductor of the CPW line, d is the bridge-to-substrate distance, t_d is the thickness of the dielectric protecting the bottom electrode, ε_r is its relative dielectric constant,

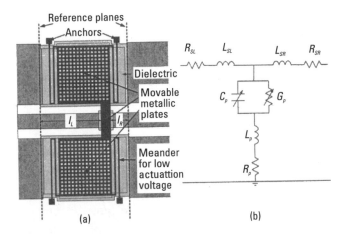

Figure 4.2 (a) Capacitive MEMS switch employed as a binary capacitor over a CPW line, and (b) binary capacitor equivalent circuit model. (*Source:* [8] ©2001 IEEE.)

and C_{fringing} is the fringing capacitance. Depending on the particulars of the implementation (e.g., Figure 4.2) and the intended application, however, it is worth pointing out that there may be other parasitic elements that would blur the simple capacitive behavior implied by (4.3), and thus they must be included in its model. For example, Peroulis et al. [8] found that for applications of the binary capacitor in filters, the extra elements in Figure 4.2(b) must be included. These elements include (1) the resistance and inductance of the bridge, denoted R_p and L_p, respectively, and obtainable from full-wave simulations; (2) the loss in the dielectric insulator coating the bottom electrode, G_p, and given by [8]

$$G_p = \begin{cases} 0 & ; \quad \text{up state} \\ \omega C_p \tan \delta & ; \quad \text{down state} \end{cases} \tag{4.4}$$

where $\tan \delta$ is the loss tangent of the dielectric; and (3) the short access lines of length l_L and l_R, as represented by their effective series resistance and inductance, R_{SL} and L_{SL}, respectively. Peroulis et al. [8] pointed out that the parasitic inductance is particularly important in the model because its values are comparable to those found in actual filters. Under the short line approximation, assumed by Peroulis et al. [8], the parasitic elements are given by

$$R_{SL} \approx 2\alpha Z_0 l_L \quad \text{and} \quad R_{SR} \approx 2\alpha Z_0 l_R \tag{4.5}$$

where α represents the line attenuation constant, usually obtained experimentally, and

$$L_{SL} \cong \frac{2Z_0}{\omega} \tan\left(\frac{\beta l_L}{2}\right) \cong \frac{Z_0 \beta l_L}{\omega} \quad \text{and}$$

$$\quad\quad\quad\quad\quad\quad\quad\quad\quad\quad\quad\quad\quad\quad\quad\quad (4.6)$$

$$L_{SR} \cong \frac{2Z_0}{\omega} \tan\left(\frac{\beta l_R}{2}\right) \cong \frac{Z_0 \beta l_R}{\omega}$$

4.2.2.2   The Binary-Weighted Capacitor Array

In the most general case, there are at least two types of tuning needs to be considered in RF/microwave circuits and systems. First, one might be interested in changing the frequency-determining parameters of a circuit/system, designed originally to operate in a given frequency band, so that it can operate at a different frequency band. An example of this would be a filter operating in a multiband wireless handset, whose response could be switched back and forth so that it could operate in two or more frequency bands. Second, one might be interested in controlling the frequency-determining parameters of a circuit/system so as to optimize its performance, which may well include switching among bands, but more interestingly would include enabling the adaptive control of the circuit/system so that it can meet a certain error function, predicated for instance, upon the real-time insertion loss, return loss, power efficiency, or harmonic level. The binary-weighted capacitor and inductor arrays [10] shown in Figures 4.3 and 4.4 were devised with these scenarios in mind.

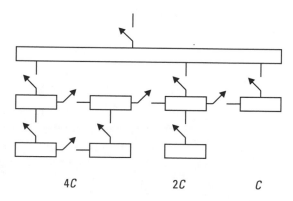

Figure 4.3   Binary-weighted capacitor array. (*After:* [10].)

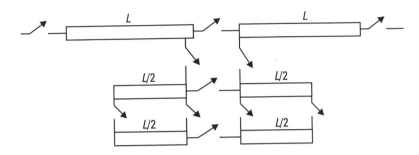

Figure 4.4 Binary-weighted inductor array. (*After:* [10].)

In the binary-weighted capacitor array, the noninvasive properties of MEM switches are exploited to reconfigure networks of elementary unit capacitor cells (Figure 4.3) so that the overall network realizes a certain capacitance. Thus, in essence, the approach discretizes the top plate area of the overall capacitor so that any value of capacitance, in steps of the unit capacitor cell value, may be set by opening or closing appropriate switches. Thus (for example, in Figure 4.3), assuming each unit capacitor cell has a value C, then to obtain an overall network capacitance of $2C$, the switches $S3$ and $S4$ would be closed, while the switches $S1$, $S2$, $S5$, $S6$, $S7$, $S8$, $S9$, and $S10$ would be open. To obtain a network capacitance of $4C$, the switches $S6$, $S7$, $S8$, $S9$, and $S10$ would be closed, while the switches $S1$, $S2$, $S3$, $S4$, and $S5$ would be open.

4.2.3 Inductors

4.2.3.1 The Binary-Weighted Inductor Array

It is well known that inductors and capacitors for operation at microwave frequencies may be realized from short sections of transmission lines (e.g., of length less than one-quarter the operating wavelength) [11]. Thus, a tunable inductor may be obtained by interconnecting, preferably via MEMS switches, network unit inductor cells so that the overall network exhibits a certain inductance. Such is the idea behind the binary-weighted inductor array [10] shown in Figure 4.4. For example, in order to obtain an inductance value of $L21$, switches $S1$, $S2$, and $S3$ would be closed, and switches $S4$, $S5$, $S6$, $S7$, $S8$, $S9$, $S10$, $S11$, and $S12$ would be open. On the other hand, if switches with an inductance of value $2L1 + 2L2$ were desired, then switches $S1$, $S3$, $S4$, $S5$, $S7$, $S10$, and $S9$ would be closed, while switches $S2$, $S6$, $S11$, and $S12$ would be open.

4.2.3.2 Series and Shunt Tunable Inductor Arrays

In addition to the systematic (binary-weighted) inductor array considered above, other arrangements for tuning series and shunt inductors, in the context of MEMS switch switching, have been advanced, as shown in Figures 4.5 to 4.7 [12]. In Figure 4.5, MEMS switches S_1, S_2, ..., S_x are connected in parallel with inductors L_1, L_2, ..., L_x, which, in turn, are connected in series between nodes IN1 and OUT1. With all switches open, the total inductance between nodes IN1 and OUT1 equals the sum of all the series-connected inductors. When any of the switches is closed, however, the low impedance of the switch bypasses that of the inductor, in essence short-circuiting it, and the total inductance decreases by that of the shorted inductor. For example, if all inductors have a common value L, then the total inductance between nodes IN1 and OUT1 may be set to be any multiple of L, from a minimum of L to a maximum of X times L (i.e., XL), where the minimum value is obtained by closing all switches but one, and the maximum value is obtained by opening all switches. If all switches are closed, a nearly zero inductance is

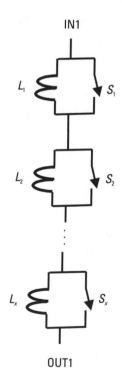

Figure 4.5 Series-connected shunt-switched inductor array and parallel-connected series-switched inductor array. (*After:* [12].)

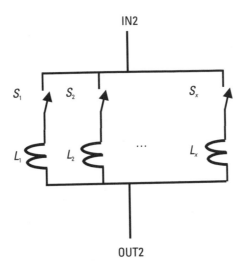

IN2

S_1 S_2 S_x

L_1 L_2 ... L_x

OUT2

Figure 4.6 Parallel-connected series-switched inductor array. (*After:* [12].)

obtained. In Figure 4.6, MEMS switches S_1, S_2, ..., S_x are connected in series with inductors L_1, L_2, ..., L_x, which, in turn, are connected in parallel between nodes IN2 and OUT2. When all switches are closed, the reciprocal of the total inductance, between nodes IN2 and OUT2, equals the sum of the reciprocals of all the parallel-connected inductors. When any of the switches is open, however, the high impedance of the switch essentially disconnects the inductor in question. For example, in a configuration of four inductors with a common value L, the total inductance will vary from a maximum of L, when all switches are open but one, to a minimum value of $L/4$, when all switches are closed. Clearly, by combining the series- and parallel-connected inductor arrays, an even more ample and fine-grained set of inductance values becomes available (Figure 4.7).

4.2.4 Tunable CPW Resonator

It is well known that one of the factors limiting the self-resonance frequency of spiral inductors is the inevitable need to use an air-bridge for connecting the inner terminal of the spiral to the output terminal, and the concomitant parasitic capacitance between the air-bridge and the underlying spiral traces. The tunable CPW resonator, demonstrated by Ketterl, Weller, and Fries [13], exploits this topological accident in the spiral inductor in order to transform it into a tunable resonator (Figure 4.8).

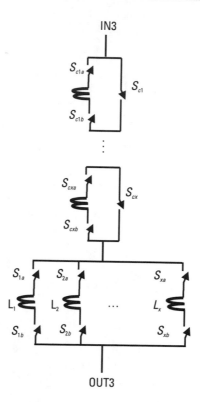

Figure 4.7 Series and parallel combination of inductor arrays. (*After:* [12].)

In the tunable resonator, the air-bridge is replaced by an electrostatically actuated cantilever beam anchored at the center of the spiral. The spiral inductor, together with the cantilever beam-to-spiral capacitance, embodies an LC resonator. When a voltage beyond pull-in is applied between the cantilever beam and the underlying SiO-coated spiral inductor, the beam "zipping" deflection action changes the beam-to-spiral capacitance, thereby changing the resonance frequency of the structure. A prototype achieved resonance tuning between 4 and 7 GHz, under biases from 0 to 40V, with corresponding Qs between 17 and 20.

4.2.5 MEMS Microswitch Arrays

So far in this chapter we have addressed techniques for the reconfigurability of more or less discrete circuit elements. Another vein in the area of

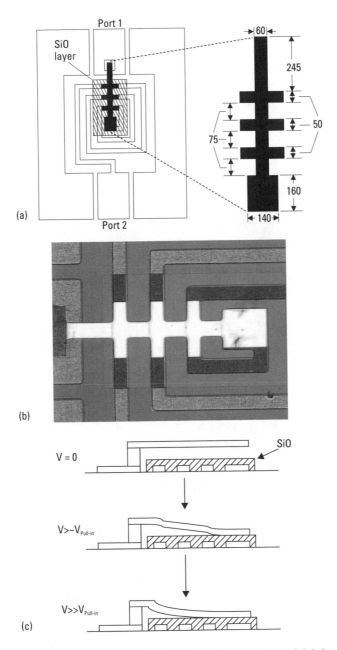

Figure 4.8 Tunable CPW resonator: (a) layout; (b) optical image of fabricated device; and (c) side view of suspended beam showing the cantilever. Top: at 0V bias; middle: after pull-in; bottom: in continual tuning state with increase in bias as the beam zips. (*Source:* [13] ©2001 IEEE. Courtesy of Prof. T. Weller and Mr. T. Ketterl, University of South Florida.)

reconfigurability—namely, the MEMS microswitch array [14]—addresses the reconfigurability of distributed microwave components (i.e., of the very metal traces, or patterns, that would otherwise define the interconnection transmission lines and tuning stubs of microstrip-based microwave circuits). The fundamental enabler of this paradigm, the microswitch, is shown in Figure 4.9. The microswitch is a cantilever beam–type structure that can be arrayed in two dimensions with an interelement pitch of 100 μm. For implementation on a fused silica substrate ($\varepsilon_r = 3.8$), this size corresponds to 1/20th the wavelength at 100 GHz or 1/200th the wavelength at 10 GHz, so that no issues of line-length quantization are elicited [14]. By addressing the two-dimensional array, where each microswitch may be thought of as a pixel, any given metal pattern image can be defined on the substrate, particularly as it is appropriate to a matching or tuning network. For example, Figure 4.10 shows the microswitch array-based tuning for a reconfigurable power amplifier, in which both the input and output matching networks are reconfigured to retune the optimum frequency of the amplifier.

4.3 Reconfigurable Circuits

Impedance matching is one of the fundamental steps in the design and production of an RF/microwave circuit [15, 16]. In low-noise amplifiers (LNAs) and power amplifiers, properly tuned input/output matching networks are

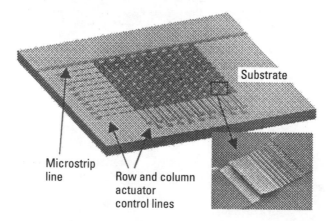

Substrate

Microstrip
line   Row and column
              actuator
              control lines

Figure 4.9 Conceptual diagram of the 2-D microswitch array. Inset: SEM of a single microswitch. The corrugations are added for mechanical strength in the cantilever arm. (*Source:* [14] ©2000 IEEE.)

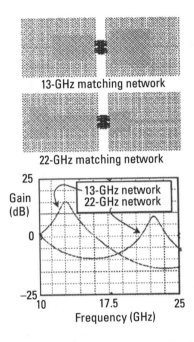

Figure 4.10 Example of a reconfigurable power amplifier. The impedance matching network at the input and output of the transistor is changed to retune the optimum operating frequency of the amplifier. (*Source:* [14] ©2000 IEEE.)

crucial to meeting required noise figure and power efficiency requirements. In the production of low-volume MICs, or hybrid (discrete) circuits, it is the rule to manually tune the circuits until the desired performance levels are met; but this activity becomes too time consuming and expensive at millimeter-wave frequencies, not to mention impractical for high-volume applications and even impossible for MMICs. Thus, there is a strong incentive to exploit the power of RF MEMS to implement classic impedance matching schemes in an automated, reconfigurable fashion.

4.3.1 Double-Stub Tuner

A case in point that pertains to reconfigurable impedance matching is the demonstration by Lange et al. [17] of a reconfigurable double-stub tuner using MEMS switches. The double-stub tuner [18] is a popular impedance matching technique in microwave circuit design. Given a load, $Y_L = G_L + jB_L$, the technique exploits the impedance-transforming properties of transmission lines to transform its real and imaginary parts, $G_L \rightarrow G_{in} = Y_0$ and $B_L \rightarrow$

$B_{in} = 0$, into a real input impedance, $Y_{in} = G_{in} = Y_0$, which represents a perfect match if Y_0 is the characteristic admittance of the system. This is accomplished via the circuit topology of Figure 4.11, whose input admittance is given by [19]

$$Y_{in} = Y_0 \times \frac{G_L + jB_L + jB_1 + jY_0 \tan \beta d}{Y_0 + (G_L + jB_L + jB) \times j \tan \beta d} + jB_2 \qquad (4.7)$$

By solving for the real and imaginary parts of (4.7) and imposing conditions on B_1, B_2, and d, so that $G_{in} = Y_0$, and $B_{in} = 0$, the matching problem is solved. In particular [17], the stub separation distance d is chosen such that

$$0 \leq G_L \leq \frac{Y_0}{\sin^2 \beta d} \qquad (4.8)$$

and B_1 and B_2 are given by

$$B_1 = -B_L \pm \frac{Y_0 + \sqrt{(1 + \tan^2 \beta d)G_L Y_0 - G_L^2 \tan^2 \beta d}}{\tan \beta d} \qquad (4.9)$$

and

$$B_2 = \frac{\pm Y_0 + \sqrt{(1 + \tan^2 \beta d)G_L Y_0 - G_L^2 \tan^2 \beta d}}{G_L \tan \beta d} + G_L Y_0 \qquad (4.10)$$

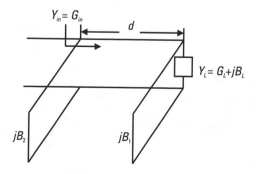

Figure 4.11 Double-stub tuner topology.

where β is the phase constant in the transmission line. Equation (4.7) imposes a limitation on the possible values of load admittance G_L that may be matched in terms of d, wavelength λ, and Y_0. For example, for $d = 0.1\lambda$ and $Y_0 = 0.02S$, GL must be less than $0.057 S$ in order to be matched to $0.02S$ (50 ohms).

The reconfigurability in the approach of Lange et al. [17] (Figure 4.12) derives from their use of RF MEMS switches to switch in and out multiple parallel-connected stubs, thus opening up the potential for realizing multiple values of B_1 and B_2. This, in turn, greatly extends the range of possible load impedance values that can be matched. With reference to Figure 4.12, then, when all switches are off, the open-circuited stubs (represented by C_{fixed}) would each adopt an equivalent capacitance given by

$$C_{eq} = \frac{C_{\text{fixed}} \times C_{\text{switch}}^{\text{off}}}{C_{\text{fixed}} + C_{\text{switch}}^{\text{off}}} \approx C_{\text{switch}}^{\text{off}} \tag{4.11}$$

if $C_{\text{switch}}^{\text{off}} \ll C_{\text{fixed}}$, thus effectively disconnecting the stub from the common node N. If the switch is on, however, then, if $C_{\text{switch}}^{\text{on}} \gg C_{\text{fixed}}$ the equivalent stub capacitance is given by

$$C_{eq} = \frac{C_{\text{fixed}} \times C_{\text{switch}}^{\text{on}}}{C_{\text{fixed}} + C_{\text{switch}}^{\text{on}}} \approx C_{\text{fixed}} \tag{4.12}$$

The stub capacitance required to fulfill (4.11) and (4.12) is a function of the frequency of operation and the values of susceptance, B_1 and B_2, and is given by [17]

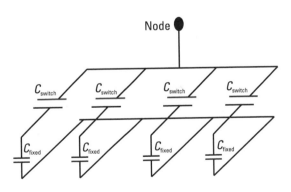

Figure 4.12 Reconfigurable stub. (*After:* [17].)

$$C_{\text{fixed}} = \frac{B}{2\pi f} \qquad (4.13)$$

For the 4-bit reconfigurable stub prototype demonstrated by Lange et al. [17], which matched loads with real and imaginary parts between 20 and 80Ω and −150 to +150Ω, respectively, the stub capacitor values ranged between 45 *fF* and 1,155 *fF*, and the switches possessed capacitances $C_{\text{switch}}^{\text{off}} \approx 35\,fF$ and $C_{\text{switch}}^{\text{on}} \approx 3\,pF$. With each 4-bit stub capable of adopting 16 susceptance values, a total of 256 configurations, or the equivalent of 256 double-stub tuner realizations, are possible—indeed, a powerful testimony to the power of RF MEMS and reconfigurability.

4.3.2   *N*th-Stub Tuner

The double-stub tuning technique described above relies on three parameters: the susceptances, B_1 and B_2, of two shunt stubs, and the distance separating them, to match a load to the working characteristic impedance of the system, usually 50Ω. Its disadvantages are that the range of loads that can be matches is limited and that it is narrowband. The triple-stub tuner system [18], in turn, can match all values of load admittances and can be optimized to increase the bandwidth. The stub-tuning concept may be generalized to that shown in Figure 4.13 [10], where a transmission line of a certain length, demarcated by its input and output ports, could have connected to it via MEM switches one or more shunt stubs of predetermined lengths at selected locations along the transmission line. By selectively closing MEM switches to some of the shunt stubs while opening others, a desired frequency response for the transmission lie can be obtained to effect impedance matching within the desired frequency range. The spacing for adjacent shunt stubs along the transmission line is about one-quarter wavelength, or an integral multiple of

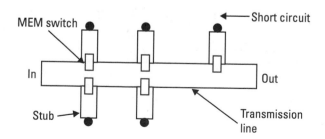

Figure 4.13   *N*th-stub/programmable transmission line tuner. (*After:* [10].)

one-quarter wavelength. The length of each shunt stub is preferably about half a wavelength, or an integral multiple of half a wavelength. However, many combinations of shunt stub lengths and spacing are also possible.

An example of an application where the *N*th-stub tuner offers an enabling function is the harmonically tunable power amplifier (Figure 4.14) [10]. Due to the high power levels produced, these power amplifiers exhibit a nonlinear transfer response characteristic that produces undesired harmonic output signals. The harmonic signals are generated at frequencies that are integral multiples of the fundamental frequency (i.e., the frequency of the input signal). With programmable stub-tuning transmission lines, to affect input and output impedance matching, a programmable stub-tuning transmission line that branches off from the output of the amplifier is added to reduce the harmonics generated by the amplifier. The harmonic tuning transmission line is connected to a plurality of shunt stubs via respective MEM switches, which are selectively turned on or off to reduce the amplitudes of harmonic signals. In general, an amplifier may produce several harmonic frequency signals that must be suppressed, and the shunt stubs with respective terminations, which may be opens or short-circuits, are arranged to prevent these signals from being transmitted to the output Out. A properly tuned harmonic tuning line segment directs the harmonic signals to a

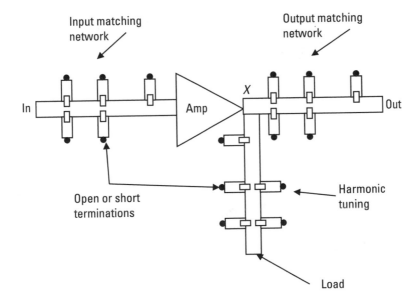

Figure 4.14 Harmonically tunable power amplifier. (*After:* [10].)

load, which absorbs the energy of the harmonic signals, so that they are not reflected back to the output line's input X. The shunt stubs with different spacing or different lengths may be used to optimize the absorption of the harmonics of interest.

4.3.3 Filters

Filters are ubiquitous building blocks in wireless systems. Indeed, a great deal of time is spent designing and redesigning filters (e.g., in satellite communications, where each satellite program must operate at a distinct set of frequencies). Reconfigurable filters, therefore, would result in more economical wireless systems, not only because fewer filters would have to be realized, but because those realized could be tuned, in principle, by computer means, as opposed to manually. Figure 4.15 shows a possible application of the binary-weighted capacitor and inductor arrays described previously to programmable filters. In the topology shown in Figure 4.15(a), an input transmission line is connected to a fix capacitor $C1$, which, in turn, is connected to a binary-weighted capacitor array $C1a$, where $C1a$ includes a plurality of

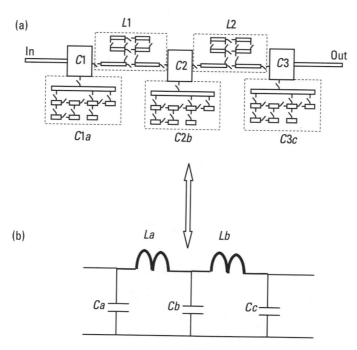

Figure 4.15 Programmable microwave filter. (*After:* [10].)

MEM switches to adjust the network's capacitance. The capacitor $C1 + C1a$, in turn, is connected to a binary-weighted inductor circuit $L1$. A filter is formed by repeating the structure of alternately connected capacitors $C2 + C2b$, $C3 + C3c$, 22a, and the inductor network $L2$. The filter's output signal is transmitted from the capacitor $C3$ to the output transmission line. Figure 4.15(b) shows the equivalent circuit of Figure 4.15(a), in which the capacitances C_a, C_b, and C_c are the sums of the capacitance values of the capacitors $C1$, $C2$, and $C3$ and the tunable capacitor networks $C1a$, $C2b$, and $C3c$, respectively. The inductances L_a and L_b are the inductance values of the tunable inductive line networks $L1$ and $L2$, respectively. Each of the capacitance values C_a, C_b, and C_c, and the inductance values L_a and L_b, which together determine the frequency response of the filter, can be changed by selectively switching at least some of the MEM switches within the capacitor and inductor networks.

4.3.4 Resonator Tuning System

Many modern wireless systems use resonator tuning schemes for changing communication frequencies. Most methods for changing communication frequencies are predicated upon coupling a voltage-controlled capacitor (varactor) to a resonator in order to change its resonance frequency. The tunable CPW resonator [13] discussed in Section 4.2.4 is an example of such a scheme, although in that case the varactor, implemented with a cantilever beam, is both the capacitor of the LC resonator and the means to effect frequency tuning. The fundamental disadvantage of varactor-coupled tuning approaches is that the intrinsic parasitic resistance of the varactor introduces losses in the resonator, thus lowering its unloaded Q. The consequence of a reduction in the unloaded Q may be appreciated by examining the carrier-to-noise (C/N) ratio in a voltage-controlled oscillator (VCO), where C/N is given by [20]

$$\frac{C}{N} = \frac{\left(2 \times Q_L \times \Delta f\right)^2 \times P_0}{\left(\text{Loss} \times f_0\right)^2 \times \left(2 \times kT \times B \times NF\right)} \qquad (4.14)$$

where Q_L is the loaded Q of the resonator, Loss is the loss factor in the resonator, f_0 is the frequency of oscillation, Δf is the offset frequency from f_0, P_0 is the output power of the oscillator, k is Boltzmann's constant, T is absolute temperature, B is the measurement bandwidth, and NF is the noise figure of the amplifier. Examination of (4.14) reveals that in order to obtain high C/N ratio, the loaded Q must be high. The loaded Q, in turn, is highest when the resonator experiences minimum external loading.

A novel technique to effect resonator tuning, which is enabled by an electrostatically actuated MEM air bridge, is indicated in Figure 4.16. In this scheme, changing the resonator's resonance frequency is accomplished by varying the capacitor or varactor coupling, rather than by varying the capacitor [21]. In essence, an interferometer, such as a Mach-Zender interferometer [19], is coupled to the resonator. Then, by way of an electrostatically actuated air-bridge disposed over one of its arms, its transmission, and consequently its coupling to the resonator, changes the resonance frequency of the resonator as described below. Using impedance-transforming properties of a transformer, the input to the primary port of a $1{:}N_t$ transformer whose secondary is loaded with a capacitor C_{tuning} or an inductor L_t results in a capacitance $N_t^2 C_{\text{tuning}}$ or an inductance L_t/N_t^2, respectively. To vary the effective coupling N_p, a Mach-Zender interferometer is coupled to dielectric resonator. In Figure 4.16, the Mach-Zender interferometer is implemented as a capacitor, specifically a ring capacitor. Thus, the Mach-Zender interferometer acts as a tunable capacitor and includes a bottom electrode, an air bridge, and a ring branch. Applying an actuation voltage V_t causes the air bridge to

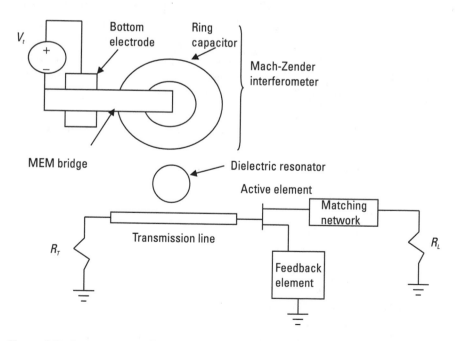

Figure 4.16 Coupling-based resonator tuning. (*After:* [21].)

deflect towards the ring branch, thus loading the ring branch with variable capacitance, which, in turn, changes the coupling to the ring branch and, as a consequence, the effective capacitance coupled to resonator. The dielectric resonator is coupled to a transmission line having a termination R_T at one end and an active element at the opposite end. The active element is coupled to both a feedback element and a matching network that is coupled to a terminating load R_L.

Referring to Figure 4.17, assuming balance amplitudes (i.e., $|t_1| = |t_2| = 1$), the transmission T, which relates the output-to-input wave amplitude ratio, is given by

$$T = |t|^2 = 2 \times \left(1 + \cos\left[L_x(k_2 - k_1)\right]\right) \tag{4.15}$$

where k is the propagation constant defined by

$$k = \omega\sqrt{L_r C_r} \tag{4.16}$$

with ω being the frequency and L_r and C_r being the inductance and capacitance per unit length, respectively. L is one-half the mean circumference of the ring. For a given L, T is a function of k_1 and k_2, and T is a measure of the coupling between input and output when there is an output transmission line. When there is not an output transmission line, the waves in each branch of the ring simply counter-propagate and T still represents the coupling to

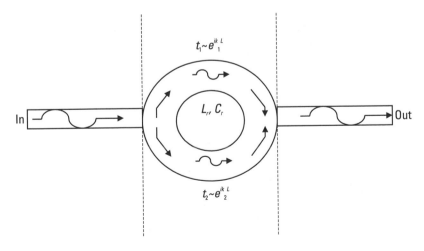

Figure 4.17 Mach-Zender interferometer/ring capacitor. (*After:* [21].)

the ring. In this case, however, it is more appropriate to consider the reflection from ring $R = 1 - T$. This coupling can be varied by changing $k_2 - k_1$, in particular, by changing C_r on one of the ring branches. Thus, the concept allows the tuning of the resonator yet without deteriorating its Q by the tuning mechanism.

4.3.5 Massively Parallel Switchable RF Front Ends

Many communications systems demand the ability to receive narrowband signals that can occur anywhere, in any one of a number of channels, within a wide frequency band. Since the equipment receiving these signals must operate in coexistence with high-power transmitters [22] and because it is imperative to avoid interference, it is required that excellent filters with narrow instantaneous bandwidth, high out-of-band rejection, wide tunability, and low insertion loss be utilized. As these requirements are only met by unrealizable filters, the usual approach to the problem involves the parallelization of the receiver into independent channels, each one containing a filter of realizable characteristics (Figure 4.18) [22]. The key to this massively parallel receiver scheme, however, is to utilize, for signal routing and reconfiguration purposes, RF/microwave switches with virtually ideal performance (particularly low insertion loss) because they directly impact the noise figure,

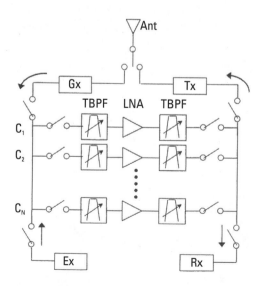

Figure 4.18 Switchable front-end operating simultaneously with other RF transmitters. (*Source:* [22]. ©1998 IEEE.)

high linearity (because in some case they must route high-power signals), and low power consumption (because of the large number of them that are required). This is precisely one of the quintessential opportunities that may be enabled by RF MEMS switches [22]. Brown [22] pointed out that traditional implementations of this architecture, based on conventional switch technology (i.e., pin diode switches), are very massive, power-consuming, and expensive. For example [22], if the pin diode switches utilized in the front-end of the ARC-210—perhaps the premier radio for military airborne communications in the VHF and UHF bands between 30 and 400 MHz—were to be replaced by MEM switches, the front-end noise figure would improve by 0.5 dB (from 4.5 to 4 dB), the transmitter-to-receiver isolation due to MEM switches in combination would improve by 20 dB (from 60 to more than 80 dB), and the total power consumption would be reduced from approximately 100 mW to less than 1 mW. Clearly, these projections could apply to the multiband/multistandard wireless transceivers that will enable the ubiquitous communications vision.

4.3.6 True Time-Delay Digital Phase Shifters

Phase shifters are at the heart of phased array antennas [23]. In simple terms, a phased array antenna consists of a set of phase shifters that control the amplitude and phase of the excitation to an array of antenna elements in order to set the beam phase front in a desired direction. While phase shifters that provide a lumped-element circuit phase shift as well as those providing a physical time delay phase shift may be employed in the implementation of a phased array, the true time delay approach enables frequency-independent beam steering, which permits the realization of phased arrays with wide instantaneous bandwidth [22]—a highly desirable feature. Consider, for example, the conventional and true time delay phased array antenna schemes depicted in Figure 4.19.

In both schemes, we find two antenna elements separated by a distance d, and driven through phase shifters in such a way that beams are set up in the direction θ_1 when the input frequency is ω_1, and θ_2 when the input frequency is ω_2. We notice, however, that whereas in Figure 4.19(a) the beam direction when the input frequency is ω_2 differs markedly from that when the frequency is θ_1, in Figure 4.19(b) the beam direction for the two frequencies is virtually identical. Let us examine this situation.

To maximize radiation in the direction θ_1, the waves emitted from the adjacent antenna elements must interfere constructively in that direction, which requires that the path length difference between these waves (namely, k_1,

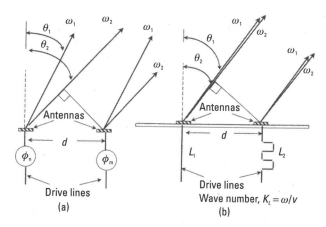

Figure 4.19 Schematic diagram of beam steering between two adjacent antenna elements driven by (a) conventional phase shifters and (b) true time delay phase shifters. (*Source:* [22] ©1998 IEEE.)

$d \sin \theta_1$) be equal to the phase difference with which the two elements are excited (namely, $\Delta\phi$). Thus, we obtain the relation $\theta_1 = \sin^{-1}(\Delta\phi \times c/\omega_1 d)$, which gives the beam direction in terms of the difference in phase of the excitation between the two elements, the frequency, and the separation. If the phase shift between elements $\Delta\phi$ varied linearly with frequency, then the ratio $\Delta\phi/\omega$ would be frequency-independent, and therefore, the beam direction would be independent from input signal bandwidth. This frequency-independence of the phase shift is difficult to achieve in lumped-element LC circuits; however, it is easy to obtain with the true time delay phase shifter approach, where $\Delta\phi = k(L_2 - L_1) = \omega(L_2 - L_1)/v$ and L and v are the physical length and the velocity of propagation in the delay lines, respectively. Inserting this value into the direction angle gives $\theta = \sin^{-1}(c(L_2 - L_1)/vd)$, which is frequency-independent.

Clearly, since with a phased array antenna one is interested in directing the beam, possibly containing broadband signals, into a multitude of directions, it is necessary to employ true time delay phase shifters with not just two, but as many as practical phase shift states. One phase shift topology, which is enabled by MEM switches, is shown in Figure 4.20.

In this digital phase shifter, the overall phase shift is set by properly configuring the switches so that the RF signal is directed through one of 2^N input-to-output path lengths and binary loop combinations. Clearly, the performance of the switches, particularly their insertion loss and isolation, is critical for the successful implementation of the scheme—and what better candidate than RF MEMS switches!

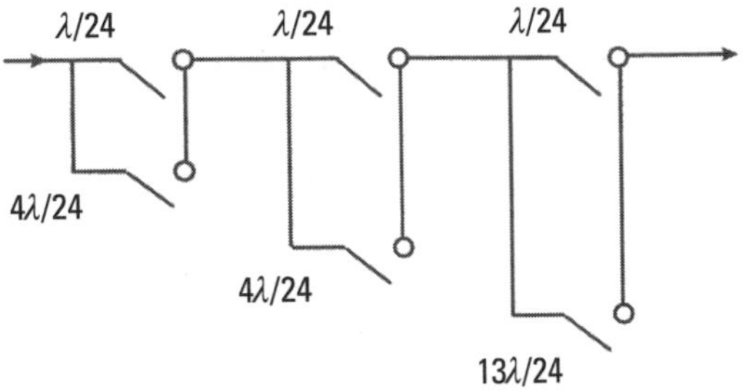

Figure 4.20 Schematic diagram of true time delay phase shifter in which three different binary loops are connected in series to provide 2^3 possible electrical delays between input and output. Extending the topology to N stages to produce 2^N possible delays is obvious. (*Source:* [21] ©1998 IEEE.)

4.4 Reconfigurable Antennas

As is well known, the radiation properties of antennas depend on the relationship between some characteristic length in their structure and the frequency being radiated. For example, dipole antennas are nearly resonant at a length close to one-half the wavelength of the excitation signal [18]. It is logical, then, that in order to increase the flexibility and usability of antennas, one would look into ways of reconfiguring their structure and dimensions.

4.4.1 Tunable Dipole Antennas

In the tunable dipole antenna [24] (Figure 4.21), a set of symmetrically located center-fed and segmented dipoles are networked via a two-dimensional array of MEM switches. Then, by closing and opening the MEM switches in an intelligent manner, the shape and length of the antenna are reconfigured and, consequently, its radiation pattern. The concept can easily be extended to a variety of antennas, such as Yagi-Uda antennas, log periodic antennas, helical antennas, and spiral plate and spiral slot antennas [25].

An extension of the tunable dipole antenna is the reconfigurable multiband microstrip resonator antenna, as shown in Figure 4.22 [26]. In this antenna, a microstrip resonator 16, on substrate 20, designed to radiate at the highest frequency (band) of interest, is excited by a signal traveling down the microstrip line, printed on substrate, through a coupling slot, defined on ground plane. The lengths of both the resonator and the coupling slot are chosen to be approximately one-half of the wavelength [26] corresponding to the highest frequency of interest. The antenna is made reconfigurable by MEM switch, which, when in the on state, changes the resonator length to

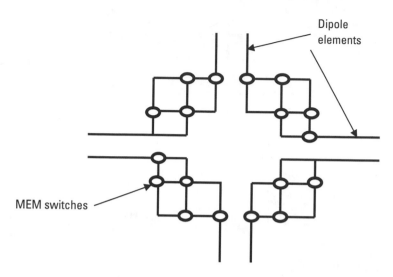

Figure 4.21 Tunable dipole antenna. (*After:* [24].)

include the additional piece of transmission line. The physically longer resonator resonates and radiates most efficiently at a lower frequency (band). To
maximize the power coupling from microstrip through slot to the resonator,
the microstrip is terminated by quarter-wave open-end tuning stubs, which
induce short-circuits at the junction of the highest and lowest frequencies of
interest with the microstrip. Then, by exploiting the high isolation of the
MEM switch in the off state, and its low insertion loss in the on state, the
resonator length can be switched between its high-frequency configuration,
main resonator element alone, and its low-frequency configuration, which
includes the extra tuning line segment. This realizes a multifrequency tunable antenna.

4.4.2   Tunable Microstrip Patch-Array Antennas

The tunable dipole antenna and the aperture-coupled microstrip line resonator antenna are made reconfigurable by changing the length of the radiating
elements, which is directly related to the wavelength at which they resonate.
Another popular radiating element, the microstrip patch, is two-dimensional
in nature and finds extensive application in antenna arrays. Interest in multifrequency patch arrays led to the reconfigurable antenna array concept [27]
(Figure 4.23).

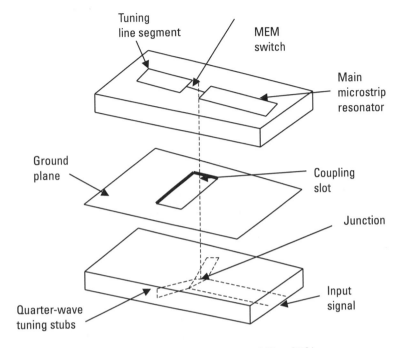

Figure 4.22 Aperture-coupled reconfigurable antenna. (*After:* [25].)

In this antenna, groups of patches are electrically connected via MEM switches. The various groups operate at multiple frequencies, according to their size. By electrically connecting or disconnecting patches, the resulting overall patch geometry acquires a side effective length and shape that is concomitant with the desired frequency of operation.

4.5 Summary

In this chapter we have presented a mostly descriptive résumé of many novel emerging RF MEMS–enabled devices, circuits, and systems. The novel devices included the resonant MEM switch, the binary MEM capacitor, the binary-weighted capacitor and inductor arrays, the tunable CPW LC resonator, and the microswitch array. The novel circuits included the reconfigurable double-stub tuner, the *N*th-stub tuner, a variety of reconfigurable filters, a resonator tuning system, as well as novel receiver and phase shifter architectures. Finally, we presented a variety of novel antenna concepts, in particular,

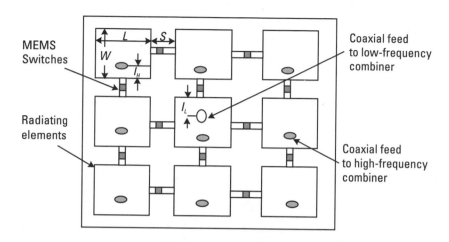

Figure 4.23 Reconfigurable microstrip array antenna. (*After:* [27].)

the tunable-dipole antenna, the slot-coupled microstrip resonator antenna, and the reconfigurable microstrip antenna. The presentation, which exposed the MEM switch, and electrostatic actuation in general, as the key elements permeating all these applications, should also serve as a resource to those individuals developing their own RF MEMS ideas and concepts. In the next chapter, we present case studies detailing the design process of RF MEMS–enabled circuits.

References

[1]   De Los Santos, H. J., *Introduction to Microelectromechanical (MEM) Microwave Systems*, Norwood, MA: Artech House, 1999.

[2]   Richards, R. J., and H. J. De Los Santos, "MEMS for RF/Wireless Applications: The Next Wave—Part I," *Microwave J.*, March 2001.

[3]   De Los Santos, H. J., and Richards, R. J., "MEMS for RF/Wireless Applications: The Next Wave—Part II," *Microwave J.*, July 2001.

[4]   Brown, E. R., "RF-MEMS Switches for Reconfigurable Integrated Circuits," *IEEE Trans. Microwave Theory Tech.*, Vol. 46, November 1998, pp. 1868–1880.

[5]   Peroulis, D. S., et al., "MEMS Devices for High Isolation Switching and Tunable Filtering," *2000 IEEE Int. Microwave Symp.*, Phoenix, AZ, 2000.

[6] Muldavin, J. B., and G. M. Rebeiz, "30 GHz Tuned MEMS Switches," *1999 IEEE Int. Microwave Symp.*, Anaheim, CA, 1999.

[7] Goldsmith, C., et al., "Characteristics of Micromachined Switches at Microwave Frequencies," *1996 IEEE Int. Microwave Symp.*, Vol. 2, pp. 1141–1144.

[8] Peroulis, D., et al., "Tunable Lumped Components with Applications to Reconfigurable MEMS Filters," *2001 IEEE Int. Microwave Symp.*, Phoenix, AZ, 2001.

[9] Pacheco, S.P., L. P. B. Katehi, and C. T.-C. Nguyen, "Design of Low Actuation Voltage RF MEMS Switch," *2000 IEEE Int. Microwave Symp.*, Boston, MA, 2000.

[10] De Los Santos, H. J., "Tunable Microwave Network Using Microelectromechanical Switches," U.S. Patent #5808527, issued September 15, 1998.

[11] Bhat, B., and S. Koul, *Stripline-Like Transmission Lines for Microwave Integrated Circuits*, New Delhi, India: John Wiley & Sons, 1989.

[12] Chang, M. C.-F., et al., "Integrated Tunable Inductance Network and Method," U.S. Patent # 5872489, issued February 16, 1999.

[13] Ketterl, T., T. Weller, and D. Fries, "A Micromachined Tunable CPW Resonator," *2001 IEEE Int. Microwave Symp.*, Phoenix, AZ, 2001.

[14] Bozler, C., et al., "MEMS Microswitch Arrays for Reconfigurable Distributed Microwave Components," *2000 IEEE Int. Microwave Symp.*, Boston, MA, 2000.

[15] Carson, R., *High-Frequency Amplifiers*, New York: John Wiley & Sons, 1975.

[16] Gonzalez, G., *Microwave Transistor Amplifiers, Analysis and Design*, Englewood Cliffs, NJ: Prentice Hall, 1984.

[17] Lange, K. L., et al., "A Reconfigurable Double-Stub Tuner Using MEMS Devices," *2001 IEEE Int. Microwave Symp.*, Phoenix, AZ, 2001.

[18] Collin, R. E., *Foundations of Microwave Engineering*, 2d ed., New York: IEEE Press, 2001.

[19] Ramo, S., J. R. Whinnery, and T. Van Duzer, *Fields and Waves on Communication Electronics*, 2d ed., New York: John Wiley & Sons, 1984.

[20] Yuen, C. M., and K. F. Tsang, "Low Voltage Circuit Design Techniques for Voltage Controlled Oscillator," *Control of Oscillations and Chaos, 1997. Proc., 1997 1st Int. Conf.*, Vol. 3, 1997, pp. 561–564.

[21] De Los Santos, H. J., "Resonator Tuning System," U.S. Patent #6,304,153 B1, issued October 16, 2001.

[22] Brown, E. R., "RF-MEMS Switches for Reconfigurable Integrated Circuits," *IEEE Trans. Microwave Theory Tech.*, Vol. 46, November 1998, pp. 1868–1880.

[23] Mailloux, R. J., "Phased Array Antenna and Technology," *Proc. IEEE*, Vol. 70, No. 3, March 1982, pp. 246–291.

[24] Lam, J. F., G. Tangonan, and R. L. Abrams, "Smart Antenna System Using Microelectromechanically Tunable Dipole Antennas and Photonic Bandgap Materials," U.S. Patent #5,541,614, issued July 30, 1996.

[25] Balanis, C. A., *Antenna Theory: Analysis and Design*, New York: John Wiley & Sons, 1982.

[26] Lynch, J., et al., "Multiband Millimeterwave Reconfigurable Antenna Using RF MEM Switches," U.S. Patent #6,069,587, issued May 30, 2000.

[27] Herd, J. S., M. Davidovitz, and H. Steyskal, "Reconfigurable Microstrip Antenna Array Geometry Which Utilizes Micro-Electro-MechanicalSystems (MEMS) Switches," U.S. Patent #6,198,438 B1.

5

RF MEMS–Based Circuit Design—Case Studies

5.1 Introduction

The design of RF MEMS–based circuits for wireless applications is predicated upon the well-established principles of RF and microwave electronics and on the novelty of RF MEMS devices. The well-established principles of RF and microwave electronics may be readily found in a variety of books [1–11], which are, no doubt, familiar to both the RF/microwave student and the practicing engineer. Given this background, the previous chapters have aimed at the following: (1) motivating the need and opportunities for RF MEMS; (2) reiterating those elements/principles of RF/microwave circuit design that are key to the successful design of RF MEMS–based circuits; (3) developing familiarity with the arsenal of novel RF MEMS circuit elements and models; and (4) introducing some of the novel RF MEMS circuits this arsenal has enabled. The aim of this chapter, then, is to present a detailed study of RF MEMS–based circuit implementations that integrates the material presented thus far. In order to do this, we have chosen, perhaps, the three most important MEMS-based circuits thus far, given their system-level impact, the way in which they are impacted by, and the way they exploit, or are drivers of, RF MEMS technology—namely, phase shifters, filters, and oscillators.

5.2   Phase Shifters

5.2.1   Phase Shifter Fundamentals

The fundamental function of a phase shifter circuit is to produce a replica of the signal applied at its input, but with a modified phase. Its performance is characterized by its insertion loss, bandwidth, power dissipation, power handling capability, and insertion phase [12]. Depending on the nature of the insertion phase (i.e., whether switchable continuously or in discrete steps), phase shifters are further classified into analog and digital, respectively.

In the analog phase shifter (Figure 5.1), the propagation properties of a transmission line, particularly its phase velocity [6],

$$v_p = \frac{1}{\sqrt{L_t C_t}} \tag{5.1}$$

where L_t and C_t are the inductance and capacitance per unit length, respectively, are varied as a function of a control voltage. Thus, in a capacitively loaded line, as the shunt load capacitance increases, the phase velocity decreases with respect to that of the unloaded line, and it takes longer for the signal to travel a certain length of transmission line. As the loading capacitors C_{Switch} introduce a periodic load of period p on the line, the phase shifter exhibits a maximum frequency of operation given by

$$f_{\text{Bragg}} = \frac{1}{\pi p \sqrt{L_t \left(C_t + C_{\text{Switch}} / p \right)}} \tag{5.2}$$

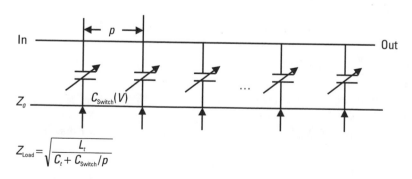

Figure 5.1 Schematic of analog phase shifter.

which is the frequency at which reflections from all impedance discontinuities introduced by the load capacitors add coherently. Thus, there is a trade-off among total line length, Bragg frequency, unloaded transmission line characteristic impedance, insertion loss, and number of switches used. One of the most recent examples of this approach is that of Barker and Rebeiz [13], who demonstrated a MEM switch-loaded 0 to 60 GHz true time delay (TTD) distributed phase shifter with 1.8-dB loss/84° at 40 GHz, and 2-dB loss/118° at 60 GHz. One of the drawbacks of analog phase shifters is that they tend to be sensitive to noise in the control voltage line.

In the digital phase shifter, the incoming input signal is routed/switched through one of many alternate paths to the output, so as to introduce specific phase shifts with minimum loss. As shown in Figure 5.2, there are four essential types of digital phase shifters: the switched line, the loaded line, the switched lowpass-/highpass-filter, and the reflection type phase shifters.

The fundamental unit (1 bit) of the switched-line phase shifter is shown in Figure 5.2(a). Switches $S1$ and $S2$, and $S3$ and $S4$, are single-pole double-throw (SPDT) switches. By their proper activation, the incoming signal is routed either through the short-length path, via $S1$ and $S3$, or through long-length path, via $S2$ and $S4$, thus endowing it with commensurate delays. The differential phase shift is given by

$$\Delta\phi = \beta\left(l_{\text{long}} - l_{\text{short}}\right) \tag{5.3}$$

where β is the phase constant.

In the loaded-line 1-bit unit [Figure 5.2(b)], switchable stubs of characteristic impedance Z_{Stub} and electrical length θ_{Stub} are separated by a spacing of line with length of θ degrees (usually $\theta = \pi/2$) and characteristic impedance Z_{01}, and disposed along a main transmission line of characteristic impedance Z_0. Each stub is terminated in switches that enable a connection to ground depending on whether they are in the on or off state. For shunt stubs exhibiting susceptances B_{OPEN} and B_{SHORT}, the differential phase shift of the 1-bit unit is given by [14]

$$\Delta\phi = \cos^{-1}\left(\cos\theta - \frac{B_{\text{OPEN}}}{Y_{01}}\sin\theta\right) - \cos^{-1}\left(\cos\theta - \frac{B_{\text{SHORT}}}{Y_{01}}\sin\theta\right) \tag{5.4}$$

The switched lowpass-/highpass-filter phase shifter unit [Figure 5.2(c)] exploits the phase shift displayed by dual lumped LC networks. For example,

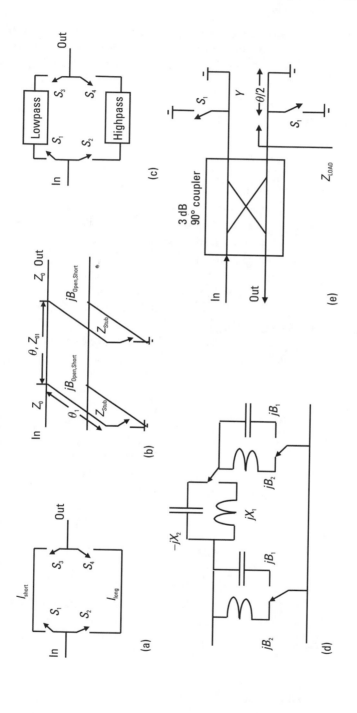

Figure 5.2 Schematic of 1-bit digital phase shifters: (a) switched line, (b) loaded line, (c) switched lowpass-/highpass-filter concept, (d) switched lowpass-/highpass-filter LC implementation, and (e) reflection. Y is the characteristic admittance of the short-terminated line of length $\theta/2$.

Davis [15] has analyzed the 1-bit unit realized by the π-network of Figure 5.2(d). Using the ABCD matrix, he obtained its transmission coefficient as

$$S_{21} = \frac{2}{2(1 - B_i X_i) + j[X_i/Z_0 + 2B_i Z_0 (2 - B_i X_i)]} \tag{5.5}$$

and the corresponding phase by

$$\theta_i = -\tan^{-1}\left[\frac{X_i/Z_0 + 2B_i Z_0 - B_i^2 Z_0 X_i}{2(1 - B_i X_i)}\right] \quad i = 1, 2 \tag{5.6}$$

The differential phase between the two states of the circuit is given by

$$\Delta\phi = \theta_1 - \theta_2 = -2\tan^{-1}\left[\frac{X_i/Z_0 + 2B_i Z_0 - B_i^2 Z_0 X_i}{2(1 - B_i X_i)}\right] \tag{5.7}$$

and the design equations, assuming unity transmission (no loss), are given as follows [15]:

$$X_i = Z_0 \tan\left(\frac{\Delta\phi}{2}\right) \tag{5.8}$$

$$B_i = \frac{1}{Z_0}\tan\left(\frac{\Delta\phi}{4}\right) \tag{5.9}$$

Finally, the reflection-type phase shifter [Figure 5.2(e)] employs a hybrid whose output port is terminated in short-circuited transmission lines of electrical length $\theta/2$. The state of the switches, whether open or closed, determines the phase of the wave coming from the out terminal with respect to that entering the in terminal. When the switches are open, the wave propagating down a transmission line of length $\theta/2$, picks up a phase shift of $\theta/2$ en route to the short-circuit termination, where it is reflected, and picks up another $\theta/2$ phase shift upon reaching the input to the line. Thus, a total phase shift of θ is achieved. Clearly, this implementation tends to save space, compared to the switched-line topology. By proper selection of the total line length and by judicious location of the switches along it, various phase shifts may be obtained. This will become clearer in Section 5.2.2. For switches with normalized admittances in the open and closed states given by $g_{OPEN} + jb_{OPEN}$ and $g_{CLOSED} + jb_{CLOSED}$, respectively, the differential phase shift is given by

$$\Delta\phi = \angle\Gamma_{OPEN} - \angle\Gamma_{CLOSED} \tag{5.10}$$

where Γ is the reflection coefficient in the open and closed switch states, respectively, given by [16]

$$\Gamma_{OPEN} = \frac{(1 - g_{OPEN}) - j(b_{OPEN} - Y\cot\theta)}{(1 - g_{OPEN}) + j(b_{OPEN} - Y\cot\theta)} \tag{5.11}$$

and

$$\Gamma_{CLOSED} = \frac{(1 - g_{CLOSED}) - j(b_{CLOSED} - Y\cot\theta)}{(1 - g_{CLOSED}) + j(b_{CLOSED} - Y\cot\theta)} \tag{5.12}$$

The motivation for using MEM switches is clear. Their negligible standby power consumption, activation power, and low insertion loss are superior to conventional semiconductor switches, thus enabling high-performance phase shifters. Recent examples of digital phase shifters are those demonstrated by Malczewsky et al. [17] and Pillans et al. [18] at X-band (8–10 GHz) and Ka-band (30–38 GHz), respectively.

An examination of the various phase shifter approaches reveals the trade-offs confronted when deciding which one to choose for a given application. For example, when size minimization is a must, as in monolithic implementations with a small available area, it is clear that the switched-line and the loaded-line topologies are prohibitive, since they require line lengths commensurate with the wavelength of interest and, thus, tend to be large at low frequencies. Similarly, implementing the lowpass-/highpass-filter type in MMIC form requires large inductors, which are problematic to realize. Lastly, we have the reflection-type phase shifter, which while requiring a hybrid whose dimension is commensurate with the wavelength of interest and, thus, may be large at low frequencies, exhibits the wideband response properties of the hybrid and uses transmission lines that only need to be half the required phase-shift length.

5.2.2   X-Band RF MEMS Phase Shifter for Phased Array Applications— Case Study

In this section we study the design of an X-band RF MEMS–based phase shifter [17]. We will proceed by discussing the specifications, the circuit design, the packaging, and the measured performance.

5.2.2.1   Specifications and Topology

The phase shifter was to be of the digital type with 4 bits of commandable phase shifts; this means, $2^4 = 16$ phase-shift states. The insertion loss was to be smaller than what is achievable with conventional technology (e.g., smaller than 7 dB), which is approximately what is exhibited by a 4-bit 6–8-GHz GaAs MMIC phase shifter [19]. X-band covers the frequency range between 8 and 10 GHz, with a wavelength of $\lambda = c/\sqrt{\varepsilon_{r\_Si}} f = 3 \times 10^8 m/s/$

$\sqrt{11.9} \times 8 \times 10^9$ Hz $\times 39.37$ in/m $= 0.428$ in, assuming silicon $\varepsilon_{r\_Si} = 11.9$ implementation; therefore, topologies that required long transmission lines (e.g., the switched-line and the loaded-line topologies) were to be avoided, not only because of size considerations but also because of the concomitant insertion losses accompanying them. In addition, long lines are incompatible with small form factor and monolithic (MMIC) integration. The switched lowpass-/highpass-filter topology was also discarded because of the large monolithic inductors it requires. The candidate of choice, then, was the reflection-type phase shifter, given its smaller size, due to requiring only half the transmission line length of the switched-line topology, and its suitability for implementation with MEM switches.

5.2.2.2   Circuit Design and Implementation

A schematic of the 4-bit reflection-type phase shifter adopted is shown in Figure 5.3. The 4-bit circuit was implemented by cascading two 2-bit stages: the first stage realizes the long states (0°, 90°, 180°, and 270°), and the second stage realizes the short states (0°, 22.5°, 45°, and 67.5°). Each core 2-bit reflection-type stage uses a Lange coupler as the 3-dB hybrid, due to its wideband property. The overall performance of the cascaded 4-bit combination produces phase shifts between 0° and 337.5° in steps of 22.5°.

The 4-bit reflection-type phase shifter design was implemented using microstrip transmission lines on 0.021-in-thick high-resistivity silicon. To minimize skin-depth effects, the conductors were made of 4-μm-thick sputtered gold ($\sigma_{Au} = 4.55 \times 10^7$ S/m) [20], which represents a thickness of almost five times the skin depth ($\delta = 0.83$ μm) at 8 GHz. It operates as follows. The

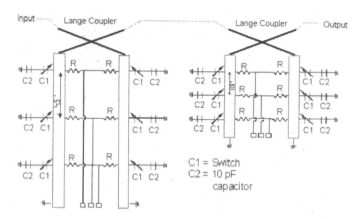

Figure 5.3 Schematic of 4-bit reflection-type phase shifter implemented by cascading two 2-bit stages. (*Source:* [17] ©1999 IEEE.)

incoming RF signal applied at the input terminal (Figure 5.3) is divided by the Lange coupler into a portion that propagates through the coupled and the direct ports down a 450-μm-wide transmission delay line. Although in this case this width realizes a characteristic impedance of 50Ω line, the line width does not have to correspond to 50Ω.

With all MEM switches off (open), the signal propagates until it reaches the end of the line, where it is reflected by the short-circuited termination. For balanced output ports (the ports to which the delay lines are connected), the reflected signals reaching them add in phase and appear at the coupler's output. When a MEM switch is on, the line is connected at a 10-pF capacitor. At 10 GHz, the 10-pF capacitor exhibits an impedance of 1.6Ω (i.e., it provides a short-circuit termination that will reflect the signal with a shorter delay than the one acquired had it traveled the total line length). Again, if the short line terminations are the same, so that the lines are balanced, the reflected signals add in phase and appear at the output of the Lange coupler.

The various phase shifts are determined by the position of the MEM switches and short-circuiting capacitors along the delay lines. The phase reference of 0° is established when the first two MEM switches, located at the beginning of the delay line, are tuned on. The 90°, 180°, and 270° phase shifts are implemented in the longer 2-bit stage by dividing the line into three 45°-long sections at 10 GHz. Thus, when the second set of MEM switches is turned on, the signal propagates 45° down to the first short-circuit termination, and 45° back to the port, for a total phase shift of 90°.

When the third set of MEM switches is turned on, the signal propagates 90° down to the short-circuit termination, and 90° back to the port, for a total phase shift of 180°. Finally, when all the MEM switches are off, the signal propagates the full 135° length of the line, to the short-circuit termination, where it is reflected and upon reaching the port acquires a phase shift of 270°.

The 22.5°, 45°, and 67.5° phase shifts are implemented by the short-length stage in a similar fashion. In these cases, however, the delay line has a width of 130 μm, commensurate with the electrodes of the MEM switches, a total length of 33.75°, and is divided into three sections of 11.25°. At 10 GHz the wavelength is approximately 0.345 in, and 11.25° is approximately 0.011 in. With the MEM switches having a total width of 0.011 in [21], it would have been impossible to realize the short phase shifts. To get around this problem, the MEM switches had to be scaled down and a small capacitor inserted in series with the delay line past all the switches, to produce a reactance that would reduce the line phase shift and, thus, make the line physically longer than 11.25°. It was pointed out in [17] that the scaling down of the switches also reduced parasitic coupling between them.

One of the advantages of the reflection phase shifter topology, which also was pointed out in [17], is that, since it required no matching networks while exploiting the wideband properties of the Lange coupler, ultra-low loss wideband performance was possible, as required by the specifications. Evidently, the transmission lines utilized in matching networks, in general, introduce inevitable ohmic, and potentially radiation and dielectric, losses, which make it difficult to achieve minimum loss. Designs for Lange couplers are actually found in the customary microwave CAD tools. Their implementation, however, requires special attention due to the large loss accompanying their large size, and because the size, in turn, sets the chip width. Also, their mismatch must be kept small (e.g., significantly less than 0.5 dB), which require the tolerance of the conductor widths and spaces to be controlled to within a few microns.

Finally, the MEM switches utilized [21], which were of the CPW membrane type, exhibited on and off capacitances of 3 pF and 35 fF, respectively, actuation voltage of about 30V, and switching times of about 5 μs.

It is noticed, upon observing Figure 5.3, that the control signal for each MEM switch is applied via resistors R and that each MEM switch is in series with a 10-pF capacitor. The resistors were selected to have a value of 10 kΩ (i.e., large enough to serve as RF-DC decoupling elements) while the capacitors embody the dual role of RF shorts for terminating the delay line and blocking capacitors for isolating the control voltage from ground.

5.2.2.3 Circuit Packaging and Performance

The assembled 4-bit MEM phase shifter circuit is shown in Figure 5.4. As observed in the circuit schematic of Figure 5.3, connections to ground play a prominent role in this design. However, the fabrication technology employed lacked vias to ground. As a result, the circuit was laid out so that the connections to ground could be implemented with wrap-around grounds at the edges of the chip. To minimize the inductance of this connection, which would have been made through 21-mil-long ribbons, small carrier plates 20-mil-thick were mounted next to the chips for grounding, resulting in grounding ribbons of only 5 mil in length. The chips, in turn, were mounted to gold-plated Kovar carriers with conductive epoxy. To exploit automated testing facilities, CPW-to-microstrip line transitions were also mounted on the carriers with epoxy and ribbon was welded to the input and output. This is indicated in Figure 5.4.

The measured performance of the circuit was as follows. The average insertion loss was 1.4 dB at 8 GHz, with a loss of 1.7 dB over greater than 30% bandwidth. The return loss was greater than 11 dB for all 16 phase shift states, and the chip-to-chip variation was within 0.3 dB.

Lessons Learned.   (1) The chip-to-chip variation at higher frequencies was caused by mismatch loss related to variations in the assembly. (2) The parasitic capacitance of the MEM switches between the switch posts and the transmission lines contributed to the phase error, which could be as high as

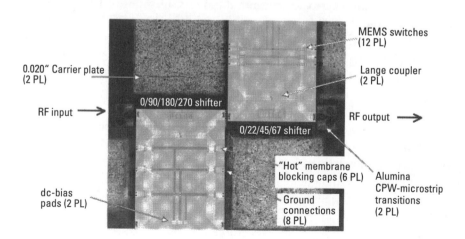

Figure 5.4 Photograph of assembled 4-bit X-band MEM phase shifter. (*Source*: [17] ©1999 IEEE.)

half the least bit, which can be lowered by shortening the line lengths. (3) The couplers were responsible for approximately 60% of the total loss. This was attributed to the lossy properties of the silicon wafer utilized; therefore, their realization in insulating substrates (e.g., alumina-based coupler) were expected to exhibit 0.2-dB lower loss. (4) The development of a technology including vias should contribute to a simpler assembly and a more compact unit.

5.2.3 Ka-Band RF MEMS Phase Shifter for Phased Array Applications— Case Study

In this section we study the design of a Ka-band RF MEMS–based phase shifter [18]. We will proceed by discussing the specifications, the circuit design, the packaging, and the measured performance.

5.2.3.1 Specifications and Topology

The phase shifter was to be of the digital type with 4 bits of commandable phase shifts; this means, $2^4 = 16$ phase-shift states. The insertion loss was to be smaller than what is achievable with conventional technology (e.g., smaller than 6.5 dB), which is approximately what is exhibited by a 4-bit Ka-band pseudomorphic high electron mobility transistor (pHEMT) MMIC phase shifter [22]. K-band covers the frequency range between 26.5 and 40 GHz, with a wavelength of $\lambda = c/\sqrt{\varepsilon_{r\_Si}} f = 3 \times 108$ m/s$/\sqrt{11.9} \times 265 \times 10^9$Hz $\times 39.37$ in/m $= 0.129$ in, at 26.5 GHz, assuming silicon $\varepsilon_{r\_Si} = 11.9$ implementation. Because of the small size afforded by this frequency range, the switched-line topology was adopted.

5.2.3.2 Circuit Design and Implementation

The implementation of this phase shifter may be understood from an examination of Figures 5.5 and 5.6.

In this implementation (Figure 5.5), 1-bit sections are utilized (Figure 5.6). Each section routes the incoming signal to one of two paths: through a reference line length, which produces a reference phase shift, or through a line with length equal to the reference line plus a length representing the desired phase shift. The switches are implemented by shunt capacitively coupled MEM switches and their function is to connect/disconnect the main transmission line to quarter-wave open stubs. As indicated in Figure 5.6, the tip of the quarter-wave stubs is a half-wave away from the T-junction between the main line and the two paths. Under these circumstances, when switch $S1$ is on (down), the open at the tip of its quarter-wave stub is

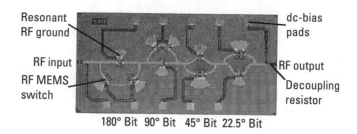

180° Bit   90° Bit   45° Bit   22.5° Bit

Figure 5.5 Photograph of 4-bit Ka-band MEMS-based phase shifter. (*Source:* [18]. ©2000 IEEE.)

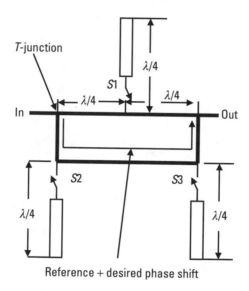

Figure 5.6 Switched-line phase shifter: 1-bit section equivalent circuit description.

transformed to an open at the *T*-junction. Therefore, the incoming signal is precluded from taking the short/reference path and, instead, propagates down the long path, which exceeds the reference length by the desired phase shift. This implements 1 bit. As shown in Figure 5.5, cascading four such sections, providing 180°, 90°, 45°, and 22.5°, results in the 4-bit phase shifter.

The MEM switches are biased by a resistive bias network connected to the membranes via open stubs. Again, 10-kΩ resistors were utilized, which simultaneously provided DC/RF isolation without slowing down the switching time of the MEM switches. The MEM switches utilized [21] were of the

CPW membrane type and exhibited on and off capacitances of 3 pF and 35 fF, respectively, actuation voltage of about 45V, and on/off switching times of about 3 to 6 μs.

The circuit was implemented using microstrip transmission lines on 0.006-in-thick high-resistivity silicon. No information was provided in [18] on metal thickness, but with a skin-depth of 0.46 μm at 26.5 GHz, conductors made of 2.5-μm-thick gold ($\sigma_{Au} = 4.55 \times 10^7$ S/m) [20], (i.e., about five times the skin depth) would have been good enough.

Upon observing Figure 5.3, it is noticed that the control signal for each MEM switch is applied via resistors R, and that each MEM switch is in series with a 10-pF capacitor. The resistors were selected to have a value of 10 kΩ (i.e., large enough to serve as RF-dc decoupling elements), while the capacitors embody the dual role of RF shorts for terminating the delay line and blocking capacitors for isolating the control voltage from ground.

5.2.3.3　Circuit Packaging and Performance

While traditional microwave packages place MMICs in a box that is hermetically sealed (to keep moisture out), these schemes require connectors and DC feed-throughs exhibiting losses at high frequencies (e.g., Ka-band), which defeats the purpose of the low-loss device inside. Therefore, the approach followed in [18] involved etching an approximately 0.007-in cavity in a lid made out of glass material and bonding the edge of the cavity to the chip using nonconductive epoxy (Figure 5.7). Glass was utilized as the lid material because of its low dielectric constant (which minimizes the perturbation or electromagnetic coupling between the lid and the circuit), its optical transparency (which allows the visual observation of the MEM switches while actuating), and its coefficient of thermal expansion (CTE) (which was close to that of silicon). The performance of the packaged phase shifter is shown in Table 5.1.

The phase shifter operated in the 30- to 38-GHz band and exhibited insertion loss between 1.8 dB at 34, for the shortest state, and 3 dB for the longest state, with an average loss of 2.25 dB. The return loss was greater than 15 dB for all states. An analysis of the losses determined that, with a line loss of 0.7 dB/cm at 34 GHz, the average loss per switch was 0.25 dB.

Lessons Learned.　The packaging of the phase shifters using the glass lid introduced two sources of loss: that due to the additional line length, concomitant with the area occupied by the lid, and that due to the transition

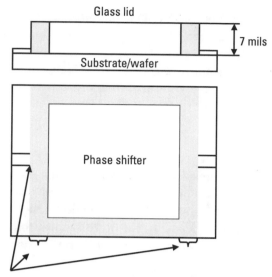

Figure 5.7   Sketch of phase shifter packaging.

Table 5.1
Performance of a 4-Bit MEMS-Based Ka-Band Phase Shifter

Phase state	0.0	22.5	45.0	90.0	180.0
Measured	0.0	10.5	32.0	92.6	172.4
Delta	0.0	−12.1	−13.0	2.6	−7.6

After: [18].

beneath the lid. The additional line length was about 0.2 cm on each side and was determined by the room necessary for proper alignment and auto-mated bonding. The accompanying loss was about 0.3 to 0.35 dB. On the other hand, the dielectric loading that results when the glass lid is placed over the microstrip using a 1- to 2-mil-thick layer of epoxy was measured to be about 0.2 dB per transition, for a total of loss contribution to the phase shifter of 0.4 dB. Thus, packaging introduced an extra loss of 0.7 to 0.75 dB. (2) The proximity of the glass lid over the phase shifter (Figure 5.7) is expected to introduce dielectric loading as well. In this case, the effect of the 7-mil glass cavity was minimal, causing insignificant changes in the inser-tion loss, and changing the resonant frequency of the phase shifter down by 0.4 GHz.

5.2.4 Ka-Band RF MEMS Phase Shifter for Radar Systems Applications— Case Study

In this section we study the design of a Ka-band RF MEMS–based true-time delay (TTD) phase shifter [23]. We will proceed by discussing the specifications, the circuit design, the packaging, and the measured performance.

5.2.4.1 Specifications and Topology

The phase shifter was to be of 4 bits with particularly wide bandwidth, as required by radar systems tasked with acquiring frequency-dependent target response and multipath signals [23]. At first sight, these specifications would suggest that an analog phase shifter of the distributed periodic capacitor load type (discussed in Section 5.2.1) might be called for. Examination of these phase shifters, however, reveals that the amount of phase shift (or time delay) they can produce is rather small. Indeed, since the phase shift is a function of the change in the bridge capacitance loading the line—and this change is preferably attained prior to pull-in—pull-in avoidance limits the working capacitance change. For instance, Kim et al. [23] point out that with the analog phase shifter the maximum delay change for an 87-ps line would only be 4 ps. The other aspect of the intended application is the wide bandwidth capability. This, as discussed in Chapter 4, may be achieved only with TTD implementations. Therefore, the topology chosen for this application was the switched-line TTD phase shifter.

5.2.4.2 Circuit Design and Implementation

The circuit design of the switched-line TTD phase shifter consists of choosing the line lengths, tailoring their bends to minimize reflections, and properly locating and impedance-matching the switches in the proximity of the T-junctions so as to maintain optimum return loss [24] across frequency even when one of the switches is in the open state and the effects of the corresponding open stub between it and the junction become manifest. Precedent for this type of design was set by Sokolov et al. [24], who positioned shunt FET switches at a distance of $\lambda_g/4$ from either the input or output junction of a 1-bit 30-GHz switched-line phase shifter. In the case of the 4-bit RF MEMS–based TTD phase shifter under study, Kim et al. [23] adopted a similar approach by optimizing the performance of a representative single bit. In particular, they began by fabricating bit number 3 as a single-bit TTD circuit. The TTD lines consisted of 55 μm wide microstrip lines on a GaAs substrate with a thickness of 75 μm, and bends made up of appropriately chamfered 90° corners to minimize reflections. The switches utilized were

Rockwell's own metal-metal contact MEMS switches, like those described in Chapter 3 [25], utilizing $100 \times 100\text{-}\mu m^2$ drive capacitors. One deviation in the signal line width from the nominal 55-μm width was dictated by the separation between the drive capacitors, which was only 80 μm; thus, the signal line had to be narrowed down to 40 μm in a 300-μm section around the junctions. The high impedance implied by this line narrowing, however, was found via simulation to be virtually inconsequential to circuit performance up to a frequency of 40 GHz. To match the open-stub transmission lines connected to the switches when in the open state, two avenues were pursued. First, in order to reduce the length of the open stub, the switch size was reduced. This resulted in an open stub with quarter-wavelength resonance frequency at 145 GHz (i.e., out of band). Second, to cancel the effective capacitance introduced by the open stub, an inductive line 16-μm-wide and 160-μm-long was introduced in series with the junctions. While, in general, it would be expected for a reactive passive element, such as the series L, shunt (open stub) C two-port embodied by the junction, to worsen the delay flatness, Kim et al. [23] point out that it need not be so. Indeed, since the group delay of such a circuit is given by

$$\tau = -\frac{d\theta}{d\omega} = \frac{d}{d\omega}\left[\tan^{-1}\left(\frac{\omega C Z_0 + \omega L/Z_0}{2 - \omega^2 LC}\right)\right] \qquad (5.13)$$

where C represents the open-stub capacitance and Z_0 is a normalizing impedance, they indicate that the ripple resulting C as a result of the second-order zero in the denominator may be minimized by properly choosing L. The design approach was validated by the reported measurements (Figure 5.8). Below 15 GHz, insertion/return loss performance of matched and unmatched circuits virtually coincides; however, beyond 15 GHz, the ripple in the return loss grows for the unmatched circuit from ~30 dB to ~10 dB, whereas for the matched circuit it remains at ~30 dB. Similarly, whereas the matched circuit exhibits less than 1-ps delay ripple up to 30 GHz, the ripple for the unmatched circuit reaches 5 ps just barely beyond 15 GHz.

5.2.4.3   Circuit Packaging and Performance

The devices were prototype vehicles tested using wafer probing techniques and were not packaged. It is clear, however, that packaging techniques and considerations similar to those employed in the phase shifter case studies previously discussed would be applicable. A photograph of the 4-bit RF MEMS TTD phase shifter is shown in Figure 5.9.

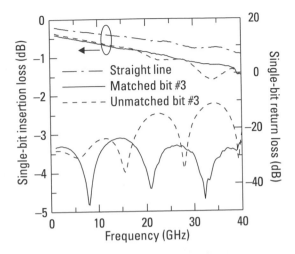

Figure 5.8 Single-bit TTD test results. Insertion loss from both the matched and unmatched delay circuits are compared with the loss from a straight 40Ω microstrip line. (*Source:* [23] ©2001 IEEE. Courtesy of Drs. R. E. Mihailovich and J. DeNatale.)

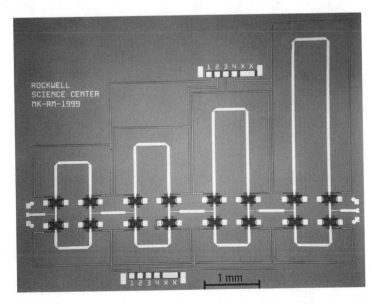

Figure 5.9 Photograph of 4-bit RF MEMS TTD phase shifter. The second longest bit (bit #3) was fabricated separately to analyze both the insertion loss and the impact of the matching section on TTD performance. (*Source:* [23] ©2001 IEEE. Courtesy of Drs. R. E. Mihailovich and J. DeNatale.)

The total area of the chip was 6×5 mm^2, and the performance was as follows. It produced delay times in the range of 106.9 to 193.9 ps in 5.8-ps increments over the dc-40 GHz frequency band. The 5.8-ps delay increment represents a phase shift of 22.5° at 10.8 GHz and is obtained by using microstrip lines with a length of 600 μm. The measured performance is shown in Figure 5.10.

An examination of the figure reveals that, in the first place [Figure 5.10(a)], the return loss for all states and over most of the frequency band remains substantially close to 20 dB, and the insertion loss variation is rather narrow. In fact, the detailed measurements of Kim et al. [23] indicate that the insertion loss for all states stayed within 0.4 dB (2.2–2.6 dB) at 10 GHz

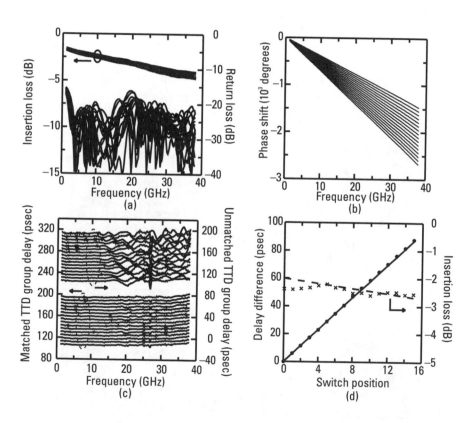

Figure 5.10 Performance of the 4-bit TTD phase shifter: (a) insertion loss and return loss; (b) phase shift from a matched TTD circuit; (c) group delay times for the matched and unmatched TTD circuit; and (d) comparison of the matched TTD phase shifter performance at 10 GHz with theory (shown as lines). (*Source:* [23] ©2001 IEEE. Courtesy of Drs. R. E. Mihailovich and J. DeNatale.)

and 0.7 dB (3.6–4.3 dB) at 30 GHz. On the other hand, the excellent phase shift linearity characteristic of the TTD circuit is observed in Figure 5.10(b), with a phase difference of 343° between the shortest and longest paths at 10.8 GHz. In terms of group delay, Figure 5.10(c) reveals that the behavior exhibited by the matched and unmatched single-bit phase shifter is also manifested in the case of the complete 4-bit circuit, particularly as it pertains to the onset of growth in group delay ripple at 15 GHz. Up to this frequency, the group delay of the unmatched circuit is flat; beyond it, the ripple increase causes the delays from different states to overlap each other [23]. At the same time, the delay exhibited by the matched TTD circuit remains essentially flat up to 40 GHz. Figure 5.10(d) shows the insertion loss and time delay information of the 4-bit TTD network for all switch (states) positions. Finally, in order to obtain maximum metal-to-metal contact and, thus, minimum insertion loss, the switch control voltage employed to set the close state was 98V.

Lessons Learned.   Detailed impedance matching is essential for obtaining good performance in switched-line TTD phase shifter circuits. Excellent insertion loss performance may be obtained at the expense of overdriving the MEMS switches in order to ensure the lowest insertion loss. Through proper design and fabrication it is possible to obtain high-yielding metal-to-metal contact MEMS switches.

5.3   Film Bulk Acoustic Wave Filters

5.3.1   FBAR Filter Fundamentals

FBAR filters are realized in two fundamental topologies: the ladder and lattice topologies [26], as shown in Figure 5.11(a, b), respectively, where the branches consist of FBAR resonators. When the coupling between resonators is realized by a shunt capacitor [Figure 5.11(c)], a symmetrical passband may be obtained if the fractional bandwidth obeys the relationship [27]

$$\frac{\Delta f}{f_0} < \frac{1}{5}\left(\frac{C_m}{C_0}\right) \qquad (5.14)$$

Since this capacitance ratio limits the bandwidth, an approach adopted by Lakin et al. [26] to increase the bandwidth involves placing an inductor L_0 in parallel with the series resonators to resonate C_0 at f_0, though at the

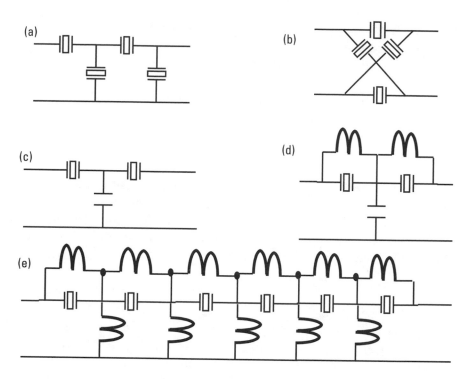

Figure 5.11 Topologies for FBAR filters: (a) ladder; (b) lattice; (c) ladder with capacitive coupling; (d) ladder with capacitive coupling and inductive widebanding; and (e) ladder with inductive coupling and widebanding.

expense of introducing a spurious passband that ranges from dc to $f_0/\sqrt{2}$ [Figure 5.11(d)]. When two-pole units are cascaded and inductively coupled [Figure 5.11(e)], the spurious passband may be moved one-half a bandwidth beyond, which results in a spurious-free passband and the movement of the spurs in the stopband to a lower level in the skirt [27].

In one approach for designing these filters, the parallel resonance introduced by the L_0–C_0 combination is assumed to be negligible near f_0, which results in a series resonator-coupled filter. Then, the usual coupled-resonator design techniques [11] may be employed, once the desired lowpass filter prototype characteristics are obtained from a filter design handbook (e.g., [28]).

While one of the key advantages of FBAR-based filters is the realization of very small-size filters [27], in comparison with conventional microwave filtering technology, the requirement for inductors to set the bandwidth, together with the lack of high-quality integrated inductors, jeopardizes such size advantage.

5.3.2 FBAR Filter for PCS Applications—Case Study

In this section we study the design of an FBAR filter developed for application in integrated single-chip radios [29].

5.3.2.1 Specifications and Topology

The filter was intended for applications in the 1.9-GHz PCS band and employed FBAR resonators in the ladder topology (Figure 5.12). In this T-cell, the filtering characteristics were realized by combining the impedance properties of FBAR resonators—namely, a low impedance, corresponding to the series resonance, near a high impedance, corresponding to the parallel resonance.

5.3.2.2 Circuit Design and Implementation

By combining series and shunt resonators of slightly different frequencies [29]—particularly two identical series-branch resonators resonating at approximately the band center, and a shunt-branch resonator resonating at the lower edge of the passband—the passband is defined as follows. The lower skirt is set by the series-resonance of the FBAR in the shunt branch; the upper skirt is set by the parallel resonance of the resonators in the series branches; and the low-insertion loss passband is achieved by the cancellation of the parallel resonance of the shunt branch and the series resonance of the series branches. The resonators themselves were of the solidly mounted type

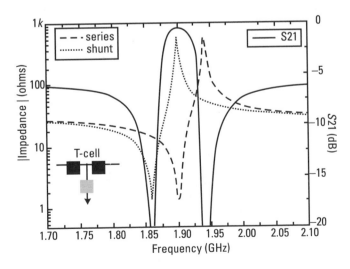

Figure 5.12 Resonator impedance and transmission for a ladder (T-cell) topology. (*Source:* [29] ©2001 Springer-Verlag. Courtesy of Dr. K. Grenier, Agere Systems.)

with an acoustic mirror of alternating layers of SiO_2 and AlN, under piezo-electric AlN.

While the topology introduced in Figure 5.12 realizes bandpass filters, the bandwidth may be limited enough to preclude meeting the specifications. To cover the desired bandwidth, external inductors in series with the FBAR resonators were employed. In this case, 1-nH inductors had to be utilized.

5.3.2.3 Circuit Packaging and Performance

While the filter under discussion was in a prototypical stage of development, similar filters for the same application band can occupy an area of 0.63 × 1.25 mm [30] and are packaged in an ordinary SO8 assembly. Figure 5.13 shows the measured performance of the PCS FBAR filter.

Lessons Learned.   Reaching the full potential for filter miniaturization will require the integration of high-quality inductors together with the FBARs.

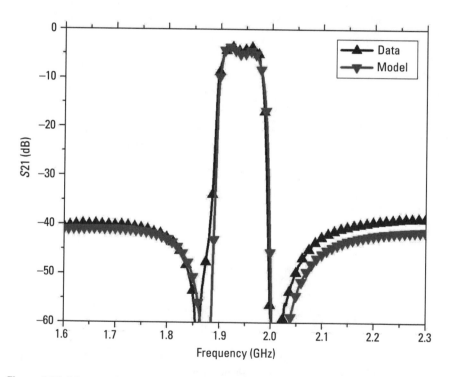

Figure 5.13 Measured performance of PCS FBAR filter made on an acoustic mirror using series inductors to widen the resonator's bandwidth. (*Source:* [29] ©2001 Springer-Verlag. Courtesy of Dr. K. Grenier, Agere Systems.)

For example, the realization of an FBAR duplexer [31] for the PCS band required external inductors in the 1–4-nH range, which were implemented by meandering them around the packaged FBAR filters on a 5.6 × 11.9–mm printed circuit board.

5.4 RF MEMS Filters

When it comes to filtering, MEMS fabrication technology has been exploited for two main purposes [32]: (1) the micromaching of structures (e.g., cavities) so small that their fabrication utilizing conventional machining techniques would be prohibitively expensive or virtually impossible due to tolerance limitations, and (2) the creation of micromechanical resonator-based filters, typically utilizing polysilicon beam resonators. As examples of the former, which is typically applied to millimeter-wave frequencies, the works of Brown and Rebeiz [33] and of Kim et al. [34] may be cited. As examples of the latter, which so far has only been applied to frequencies lower than 200 MHz, the works of Bannon, Clark, and Nguyen [35] and of Wang and Nguyen [36] may be cited. In what follows, we present two case studies: one being an example of a tunable micromachined filter, and the other an example of a micromechanical MEMS filter.

5.4.1 A Ka-Band Millimeter-Wave Micromachined Tunable Filter—Case Study

The motivations for developing tunable filters might be traced to three factors: cost, weight, and power dissipation. In modern satellite communications, for instance, where an enormous number of filters are employed, there is extreme interest in minimizing both manufacturing cost, weight, and power consumption. The potential of MEMS technology to permit the batch fabrication of tunable filters is rather appealing, as these might in principle obviate the need for expensive manual tuning in favor of electronic tuning during system integration and alignment, thus reducing cost. On the other hand, the availability of inexpensive tunable filters opens up new opportunities, not only for novel systems architectures, but, by increasing filter functionality, thus reducing the number of individual filters needed, which reduces overall weight. Finally, since MEMS avails itself of electrostatic schemes for tuning, the power dissipation incurred in implementing tunability is negligible.

In this section we study the cases of millimeter-wave micromachined tunable filters demonstrated by Kim et al. [34].

5.4.1.1 Specifications and Topology

Two filters intended for application in highly integrated transmitters and receivers utilized in millimeter-wave multiband communication systems were implemented: a lumped-element version with a 4.7% bandwidth, centered at 26.8 GHz, and a coupled-resonator version with a 8.5% bandwidth, centered at 30.6 GHz. The filter topologies are shown in Figure 5.14.

In the lumped-element version [Figure 5.14(a)], the LC resonators were implemented as π-networks with series spiral inductors and shunt capacitors implemented with micromachined cantilever-type varactors. In the coupled distributed resonator version [Figure 5.14(b)], the resonators were implemented by half-wavelength resonators terminated in varactor connected to ground. Thus, by varying the capacitance of the varactors, the effective electrical length of the resonators deviated from half-wavelength and tuning is affected.

5.4.1.2 Circuit Design and Implementation

The filters were implemented in grounded coplanar waveguide (GCPW) media (Figure 5.15), and their design was carried out via a full-wave electromagnetic simulator [34]. For the lumped-element realization, the simulation entailed varying the inductive (magnetic) coupling between the spiral inductors of the LC resonators, which was achieved by optimizing the filter response as a function of the separation between the inductors. The half-wavelength resonator-coupled filter version can be designed using coupler synthesis tools and fine-tuned with the full-wave solver. In addition, in order

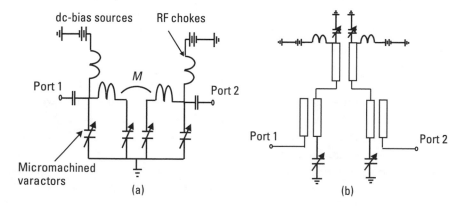

Figure 5.14 Topologies for MEM tunable filters: (a) two-pole lumped-element filter; and (b) two-resonator-coupled filter. (*Source:* [34] ©1999 IEEE.)

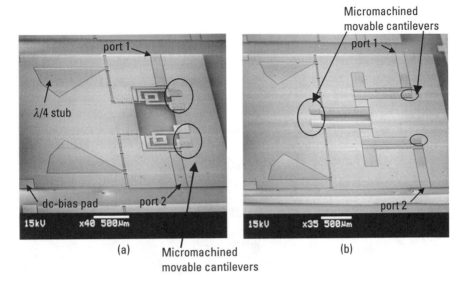

Figure 5.15 Microphotograph of two micromachined tunable filters: (a) two-pole lumped-element filter (3.4 × 2.9 mm); and (b) two-pole coupled distributed resonator filter (4.1 × 3.4 mm). (*Source:* [34] ©1999 IEEE. Courtesy of Dr. Y. Kwon, Seoul National University.)

to assess attainable tuning ranges, simulations were conducted while setting the varactor parallel plate gap to various distances: 6 μm, 5 μm, and 4 μm. It is interesting to note that, in this simulation, the lumped-element version exhibited a downward shift of 6.4% in center frequency with a capacitor gap reduction of 2 μm, while the distributed resonator version exhibited a downward shift of 2.6% for a varactor gap reduction of 1 μm. Also shown in Figure 5.14 are RF chokes to permit the noninvasive application of the varactor control voltage.

5.4.1.3 Circuit Packaging and Performance

The devices were research demonstration vehicles tested using wafer probing techniques and were not packaged. Given the presence of MEM varactors, however, it is clear that packaging techniques and considerations similar to those employed in the case of phase shifters would be applicable. Figure 5.16 shows the measured performance [34].

The lumped-element filter exhibited a minimum insertion loss of 4.9 dB, a return loss of ~10 dB, and a tuning range of 4.2% over a voltage range of 65V. The coupled distributed resonator filter, on the other hand, exhibited a minimum insertion loss of 3.8 dB, an average return loss of ~15 dB,

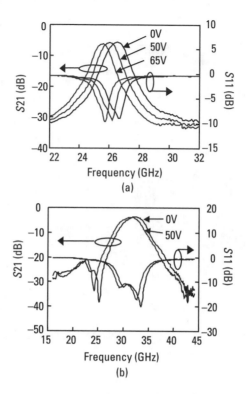

Figure 5.16 Measured responses of micromachined tunable filters: (a) two-pole lumped-element filter; and (b) two-pole coupled distributed resonator filter. (*Source:* [34] ©1999 IEEE. Courtesy of Dr. Y. Kwon, Seoul National University.)

and a tuning range of 2.5% over a voltage range of 50V. In comparing the observed tuning range with the simulated one, Kim et al. [34] point out that the smaller measured tuning range results from the lower capacitance change of a partially deflected beam (i.e., larger deflection at the beam tip than at the anchor), whereas the simulated change assumed uniform gap shrinking.

Lessons Learned. The millimeter-wave tunable filters of this case study are an excellent example of the potential of RF MEMS—that is, functionality at low cost and power consumption. While not addressed here, much work remains to be done to perfect tunable filters. For instance, it is obvious that the tuning range of cantilever-type varactors is limited by pull-in. Also, in some applications it might be necessary to affect tuning at some prescribed time rate. The performance of high-frequency filters is extremely dependent on stray coupling; thus, there is no guarantee of what response will ultimately be obtained unless the packaged device is tested.

5.4.2   A High-Q 8-MHz MEM Resonator Filter—Case Study

The motivation for developing micromechanical resonator-based filters is readily apparent once awareness is gained of their potential to exhibit very high on-chip Qs (e.g., more than 80,000 under vacuum conditions) [37]. Traditionally, high-Q filters, which are employed in the RF and intermediate frequency (IF) strips of superheterodyne receivers, must be implemented with off-chip crystal, ceramic, or surface acoustic wave resonators. With continued miniaturization of wireless appliances, however, not only do these off-chip components introduce a lower bound to miniaturization, but attempts to obviate them has led to alternate architectures, such as direct-conversion, wideband-IF, or direct sampling, whose performance does not match that of the superheterodyne [35]. In light of this situation, micromechanical resonator-based filters have been the subject of extensive interest and exploration. While not particularly difficult, the design of micromechanical filters does require RF/microwave engineers to engage in the detailed aspects of mechanical design, in addition to the usual electrical aspects. Thus, this case study provides an exercise on dual-domain (i.e., RF and MEMS) circuit design.

5.4.2.1   Specifications and Topology

The filter prototype intended for operation around 8 MHz is representative of those found in the high-frequency 3–30-MHz frequency range, which is typical of applications such as cellular telephony [10]. The filter topology [Figure 5.17(a)] consists of two identical clamped-clamped MEM resonators, coupled by a flexural-mode beam, together with biasing, input and output driving and sensing interfaces. The equivalent mechanical circuit is shown in Figure 5.17(b) and this, together with the mechanical-to-electrical analogies presented in Chapter 3, facilitate the design in the electrical domain.

5.4.2.2   Circuit Design and Implementation

The design of MEM filters proceeds along the lines of conventional resonator-coupled filters (i.e., it hinges upon obtaining coupling coefficients that will enable a given response to be met) [32]. This entails inspection of families of normalized filter responses that are cataloged in standard

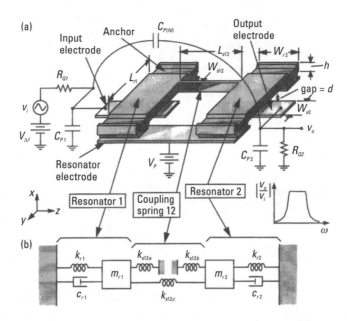

Figure 5.17 (a) Perspective view schematic of a two-resonator MEM filter, along with the preferred bias, excitation, and sensing circuitry. Significant parasitic elements are also shown in gray. (b) The equivalent mechanical circuit. (*Source:* [35] ©2000 IEEE.)

handbooks (e.g., in the "Tables of k and q Values") [28]. Once the filter degree that will meet the desired selectivity is obtained, the point of entry into these tables is the normalized resonator quality factor q_0, defined by

$$q_0 = \frac{\Delta f}{f_c} Q_0 \qquad (5.15)$$

where Δf is the bandwidth, f_c the center frequency, and Q_0 the intrinsic resonator quality factor of the resonators, assumed to be uniform. The normalized coupling coefficients and quality factors, $k_{i,j}$ and q_i, respectively, can subsequently be read off the tables. From these, the denormalized coupling coefficients and quality factors are obtained according to

$$K_{i,j} = k_{i,j} \frac{\Delta f}{f_c} \qquad (5.16)$$

and

$$Q_i = q_i \frac{f_c}{\Delta f} \tag{5.17}$$

The design of the MEM filter thus consists of determining the physical parameters of the MEM resonators and the coupling spring, the point of coupling/attachment of the coupling spring to the resonators, and the resonator coupling to the input/output [32] that will realize the above values. In the specific example under study, Bannon et al. [35] utilized the following two-resonator filter design procedure. In the first place, they chose appropriate resonator geometries to meet the desired resonance frequency, as well as proper electrode-to-resonator transducer capacitive overlap to enable sufficient coupling to input and output. Second, they chose the coupling beam width W_s with particular regards to it being manufacturable, and the beam length to realize a quarter wavelength at the center frequency. Third, they explored and determined the appropriate beam-to-resonator coupling location to achieve the desired filter bandwidth. Finally, they generated a complete electrical equivalent circuit of the filter and analyzed its performance in a circuit simulator. The details of these steps are now discussed.

Micromechanical Resonator Design

The design equations for the MEM resonator relate the geometry, material parameters, and polarization voltage required to achieve a given resonance frequency. Those for the clamped-clamped resonator were given in Chapter 3 and are reproduced below for convenience.

Nominal resonance frequency:

$$f_{r,\text{nom}} = 1.03\kappa\sqrt{\frac{\mathrm{E}}{\rho}} \times \frac{h}{L_r^2} \tag{5.18}$$

Ratio of electrical-to-mechanical stiffness:

$$g(d, V_p) = \int_{L_{e1}}^{L_{e2}} \frac{dk_e(y')}{k_m} \tag{5.19}$$

where

$$dk_e(y') = V_P^2 \frac{\varepsilon_0 W_r \, dy'}{\left(d(y')\right)^3} \quad (5.20)$$

$$d(y) = d_0 - \frac{1}{2} V_P^2 \varepsilon_0 W_r \int_{L_{e1}}^{L_{e2}} \frac{1}{k_m(y')\left(d(y')\right)^2} \frac{X_{\text{static}}(y)}{X_{\text{static}}(y')} dy' \quad (5.21)$$

$$X_{\text{mode}}(y) = \zeta(\cos ky - \cosh ky) + (\sin ky - \sinh ky) \quad (5.22)$$

with $k = 4.730/L_r$ and $\zeta = -1.101781$.

Corrected resonance frequency accounting for electrical stiffness:

$$f_0 = f_{r,\text{nom}} \sqrt{1 - \frac{k_e}{k_m}} = f_{r,\text{nom}} \sqrt{1 - g(d, V_P)} \quad (5.23)$$

The coupling of the end resonators to the input/output terminals of the filter plays an important role in determining the overall filter response's flatness. In particular, since too high a resonator Q leads to a response exhibiting isolated peaks (dissipation), lowering the Q and setting up the flatness is introduced via terminating resistors. The value of the resistors in question is given by [35]

$$R_{Qi} = \left(\frac{Q}{q_i Q_{\text{fltr}}} - 1\right) R_{xi} \quad (5.24)$$

where Q is the individual resonator unloaded quality factor, $Q_{\text{fltr}} = f_0/B$ is the overall filter quality factor, q_i is the normalized quality factor obtained during the filter synthesis procedure from a cookbook, and i designates the end resonator of interest. When the individual resonator Q is much greater than the filter Q (i.e., $Q \gg Q_{\text{fltr}}$), which is often the case, the terminating resistor is given by

$$R_{Qi} \cong \frac{\sqrt{k_{rie} m_{rie}}}{q_i Q_{\text{fltr}} \eta_e^2} = \frac{k_{rie}}{\omega_0 q_i Q_{\text{fltr}} \eta_e^2} \quad (5.25)$$

where Bannon et al. [35] point out that achieving a desired value of R_{Qi} is easiest by adjusting the electromechanical coupling factor η_e. This, in turn, dictates the beam-to-electrode gap and overlap, W_e.

Coupling Beam Design

In a mechanical sense, the coupling beam communicates the vibration from the excited resonator to the output resonator. Electrically, the role of the coupling beam is to realize the coupling coefficient K_{ij} that will produce the prescribed response shape (e.g., Butterworth, Chebyshev, elliptic, and so on). For given filter bandwidth and center frequencies f_0 and B, resonator stiffness k_{rc} and desired electrical coupling coefficient k_{12}, the stiffness of the coupling spring k_{s12} is given by [28, 38]

$$k_{s12} = k_{rc}\left(\frac{B}{f_0}\right)k_{12} \qquad (5.26)$$

Bannon et al. [35] stressed a number of key practical issues pertaining to the actual realization of the coupling spring and its effect on the resulting filter response. In the first place, the mass of the coupling springs, compared to that of the resonators, is not negligible. The effect of this is that, in essence, the mass of the resonators is increased by the spring loading, thus causing a shift in their resonance frequency and, consequently, in the overall filter response. Figure 5.18 illustrates the situation when each resonator behaves as if it has acquired half the coupling spring mass.

In the second place, the length of the coupling springs, compared to the operating acoustic wavelength, may not be negligible. In this case, the coupling spring is more accurately described by a distributed structure (i.e., an acoustic transmission line). Since a transmission line presents at its input a transformation of the impedance loading it at its output, the amount of spring mass and stiffness perceived by the resonators it couples will be a

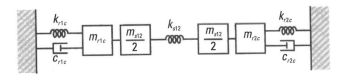

Figure 5.18 Equivalent mechanical circuit for the two-resonator filter using a coupling beam of length less than an eighth of a wavelength of the operating frequency. (*Source:* [35] ©2000 IEEE.)

function of both beam dimensions and operating frequency. Thus, in general, the coupling beam must be modeled as shown in Figure 5.19.

The equivalent electrical network for the coupling spring was originally determined by Konno and Nakamura [39] (Figure 5.1) and is described by the following matrix equation when the rotation at the points of contact may be neglected:

$$
\begin{bmatrix} f_1 \\ \dot{x}_1 \end{bmatrix} = \begin{bmatrix} \dfrac{H_6}{H_7} & -\dfrac{2EI_s\alpha^2 H_1}{j\omega L_s^3 H_7} \\[2ex] -\dfrac{j\omega L_s^3 H_3}{EI_s\alpha^3 H_7} & \dfrac{H_6}{H_7} \end{bmatrix} \begin{bmatrix} f_2 \\ \dot{x}_2 \end{bmatrix} = \begin{bmatrix} A & B \\ C & D \end{bmatrix} = \begin{bmatrix} f_2 \\ \dot{x}_2 \end{bmatrix} \tag{5.27}
$$

where

$$H_1 = \sinh\alpha\sin\alpha \tag{5.28}$$

$$H_3 = \cosh\alpha\cos\alpha - 1 \tag{5.29}$$

$$H_6 = \sinh\alpha\cos\alpha + \cosh\alpha\sin\alpha \tag{5.30}$$

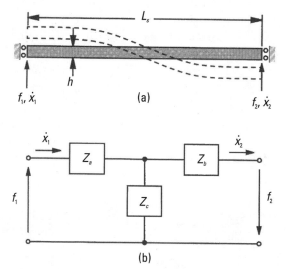

Figure 5.19 (a) Coupling beam under forces f_1 and f_2 with corresponding velocity responses; and (b) general transmission line T-model for the coupling beam. (*Source:* [35] ©2000 IEEE.)

$$H_7 = \sin\alpha + \sinh\alpha \tag{5.31}$$

with W_s being the beam and h its thickness, and $\alpha = L_s (\rho W_s h \omega^2)/((EI_s))^{0.25}$, and $I_s = W_s h^3/12$. Equation (5.27) is reminiscent of the ABCD or chain matrix. Therefore, the impedances are given by

$$Z_a = Z_b = \frac{A-1}{C} = \frac{jEI_s\alpha^3\left(H_6 - H_7\right)}{\omega L_s^3 H_3} \tag{5.32}$$

and

$$Z_c = \frac{1}{C} = \frac{jEI_s\alpha^3 H_7}{\omega L_s^3 H_3} \tag{5.33}$$

In order to minimize the sensitivity to beam variations stemming from layout or fabrication tolerances, Bannon et al. [35] recommend that the coupling spring length be chosen to be a quarter-wavelength of the center frequency of the filter. This can be achieved, they point out, by forcing Z_a and Z_c to adopt equal and opposite signs, which is obtained when H_6 is made zero; that is,

$$H_6 = \sinh\alpha\cos\alpha + \cosh\alpha\sin\alpha = 0 \tag{5.34}$$

From the value of α satisfying (5.34), the value for L_s is obtained, given technology-determined values for W_s and h. In this case, then, the impedances in the T-model for the coupling spring become

$$Z_a = Z_b = \frac{EI_s\alpha^3 H_7}{j\omega L_s^3 H_3} = \frac{k_{sa}}{j\omega} \tag{5.35}$$

and

$$Z_c = -\frac{EI_s\alpha^3 H_7}{j\omega L_s^3 H_3} = \frac{k_{sc}}{j\omega} \tag{5.36}$$

Substituting (5.29) and (5.31) into these equations allows one to obtain an expression for the stiffness of the quarter-wavelength beam as

$$k_{sc} = -k_{sa} = -\frac{EI_s \alpha(\sin\alpha + \sinh\alpha)}{L_s^3(\cos\alpha\cosh\alpha - 1)} \qquad (5.37)$$

A simplified electrical-domain model for the two-resonator filter is shown in Figure 5.20.

Bannon et al. [35] point out that an extra advantage of using a quarter-wavelength coupling beam is that this allows for identical resonators to be utilized, with the concomitant consequence that filter realization is easier due to the greater ability of planar processes to achieve relative matching tolerances, as opposed to absolute tolerances, which would be required if one had to produce resonators with (different) specific frequencies.

Coupling Location

Having designed the resonators and the coupling beam, it is necessary to determine the optimum location along the resonators to which the coupling beam must be attached. There are two guiding principles to determine this. First, the filter Q must be satisfied; that is,

$$k_{s12} = k_{rc}\left(\frac{B}{f_0}\right)k_{12} \rightarrow Q_{fltr} = k_{12}\left(\frac{k_{rc}}{k_{s12}}\right) \qquad (5.38)$$

Second, in micromechanical filters with similar-sized resonators and coupling beams, the ratio k_{rc}/k_{s12} cannot be varied arbitrarily—in fact, it is limited [35]. Towards this end, simulations performed by Bannon et al. [35] (Figure 5.21) in which the filter Q and its reciprocal, the percent bandwidth, were calculated as a function of coupling beam location, revealed that at

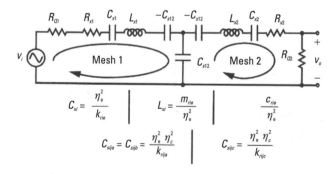

Figure 5.20 Simplified equivalent circuit for the MEM filter of Figure 5.17 using a quarter-wavelength coupling beam. (*Source:* [35] ©2000 IEEE.)

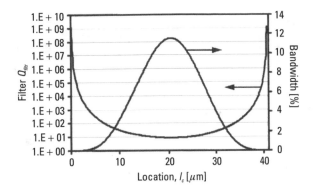

Figure 5.21 Plot of simulated Q_{fltr} and percent bandwidth versus coupling location l_c along the length of a resonator beam. (*Source:* [35] ©2000 IEEE.)

resonance, the stiffness of a clampled-clamped k_{rc} beam is larger at locations close to the anchors (i.e., at locations of low velocity). This feature, in turn, represents an additional degree of freedom in this type of MEM filter design—namely, the resonators and coupling beams may be designed independently from the bandwidth, and the bandwidth may be set by setting up the resonator-to-coupling beam point of attachment [35].

5.4.2.3   MEM Filter Design, Implementation, and Performance

For detailed design purposes, the complete equivalent circuit employed for the two-resonator MEM filter shown in Figure 5.22 should, in principle, be utilized. At the present time, however, the state of maturity of such circuit models (i.e., the lack of accurately predictable parameter values) only renders them amenable for use a posteriori (i.e., for analysis purposes). Thus, current design practice avails itself of a combination of semiempirical, numerically simulated, and theoretically approximated parameters. With this in mind, Table 5.2 shows a summary of the corresponding values that apply to the 7.81-MHz MEM filter shown in Figure 5.23(a), when adjusted to match the measured transmission characteristics shown in Figure 5.23(b). Table 5.3 contains the circuit model values extracted from fitting the model of Figure 5.22 to the measured data.

Examination of Figure 5.23(b) and Table 5.2 reveals that, while good agreement was obtained between simulated and measured transmission characteristics, this entailed assigning values to some parameters that substantially deviated from their design values. For example, the resonator Q, the beam-to-substrate gap, and the coupling location were all assigned substantially

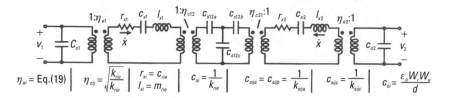

Figure 5.22 Complete equivalent circuit for the MEM filter of Figure 5.17, modeling both quarter-wavelength coupling beam design and low-velocity coupling location. Expressions for the elements are also included. (*Source:* [35] © 2000 IEEE.)

Table 5.2
HF 7.81-MHz MEM Filter Summary

Parameters	Value [1, 2, 3, 4]		Units
	Designed/ Measured	**Simulated**	
Coupling location, l_c	4.08	4.48	μm
Coupling velocity, v_c	0.12 vmax	**0.14 vmax**	M/s
Center frequency, f_0	7.81	7.81	MHz
Frequency modification factor, κ	(0.87915) [0.9]	0.87915	—
Bandwidth, B	18	18	KHz
Percent bandwidth, (B/f_0)	0.23	0.23	%
Passband ripple, PR	1.5	1.5 [0.5]	dB
Insertion loss, IL	1.8	1.8 [1.35]	dB
20-dB shape factor	2.31	[2.54]	—
Stopband rejection, SR	35	—	dB
Spur-free dynamic range, $SFDR$	~78	~78	dB
Resonator, Q	8,000	**6,000**	—
Structural layer thickness, h	1.9	1.9	μm
Resonator beam length, L_r	40.8	40.8	μm
Resonator beam width, W_r	8	8	μm
Coupling beam length, L_{s12}	20.35	20.35	μm
Needed L_{s12} for $\lambda/4$	(22.47)	—	μm
Coupling beam width, W_{s12}	0.75	0.75	μm
Coupling beam stiffness, k_{s12a}	(−82.8)	−82.8	N/m
Coupling beam stiffness, k_{s12c}	(113.4)	113.4	N/m

Table 5.2 (continued)

Parameters	Value [1, 2, 3, 4]		Units
	Designed/ Measured	**Simulated**	
Resonator mass @ I/O, m_{re}	(5.66×10^{-13})	5.66×10^{-13}	Kg
Resonator stiffness @ I/O, k_{re}	(1,362)	1,362	N/m
Resonator mass @ I_c, m_{rc}	(3.99×10^{-11})	2.84×10^{-11}	Kg
Resonator stiffness @ I_c, k_{rc}	(96,061)	68,319	N/m
Integrated resonator stiffness, k_{reff}	(1,434)	1,434	N/m
Young's Modulus, E	150	150	GPa
Density of polysilicon, ρ	2,300	2,300	Kg/m^3
Electrode-to-resonator gap, d_0	1,300	1,300	Å
Gap d_0 adjusted for depletion	1,985	1,985	Å
Electrode width, W_e	20	20	μm
Filter dc bias, V_P	35	35	V
Frequency pull-in voltage, $V_{\Delta f}$	0.12	0	V
Q-control resistors, R_{Qi}	12.2 (19.6)	**14.5** [19.6]	KΩ

1. Numbers in ( ) indicate calculated or semiempirical values.

2. Bold-faced numbers indicate significant deviations needed to match simulated curves with measured curves.

3. Numbers in [ ] indicate values expected from an ideal simulation with no parasitics and perfect termination. The value for k in the Designed/Measured column was obtained via finite-element simulation using ANSYS.

4. Top 11 rows represent simulation outputs; the rest are used as inputs for simulation.

After: [35].

different values than anticipated. To justify these discrepancies, somewhat plausible arguments were advanced by Bannon et al. [35], such as coming across a resonator with an out-of-family lower Q, a manifestation of the depletion region induced in the resonator due to the polarization voltage resonator, and a manifestation of the finite width of the coupling beam, which makes somewhat imprecise the actual location of mechanical coupling, respectively.

In terms of the actual fabricated device, in order to compensate for the fact that the two resonators did not exhibit identical resonance frequencies

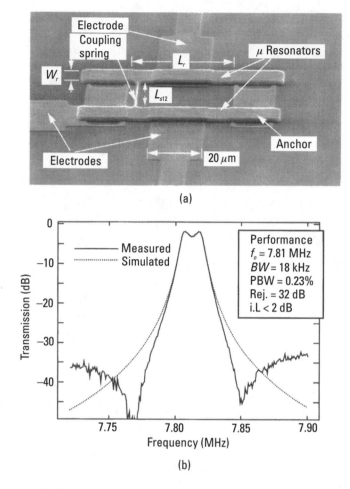

(a)

(b)

Figure 5.23 (a) SEM of a fabricated 7.81-MHz two-resonator MEM filter; and (b) Measured transmission for a terminated 7.81-MHz MEM filter with excessive input/output shunt capacitance. Here, $Q_{fltr} = 435$. (*Source:* [35] ©2000 IEEE.)

upon turn-on, it was necessary to apply a dc tuning voltage $V_{\Delta f}$ of 0.12V (frequency pulling voltage) (see Figure 5.17) to resonator 1 in order to match its resonance frequency to that of resonator 2 and, thus, obtain the correct passband in Figure 5.23(a).

In order to obtain the above data, the filter and all the necessary testing accessories (e.g., test fixtures) were placed in a custom-built vacuum chamber; only coax cables via feed-throughs provided a connection to external instrumentation.

Table 5.3
HF MEM Filter Circuit Element Values

Parameter	Value	Unit
Coupling location, l_c	4.48	μm
$C_{o1} = C_{o2}$	7.14	fF
$l_{x1} = l_{x2}$	5.66×10^{-13}	H
$C_{x1} = C_{x2}$	0.000734	F
$r_{x1} = r_{x2}$	4.62×10^{-9}	Ω
$C_{s12a} = C_{s12b}$	−0.0121	F
C_{s12c}	0.00882	F
$\eta_{e1} = \eta_{e2}$	1.20×10^{-6}	C/m
$\eta_{c12} = \eta_{c21}$	7.08	C/m

After: [35].

Lessons Learned. (1) While great strides have been made in the development of MEM filters, the art is still highly empirical. (2) This high degree of empiricism, however, might well be dealt with via the implementation of on-chip trimming techniques. (3) It appears, therefore, that to accelerate the insertion of these filters, the endeavor of developing on-chip trimming, as well as improving device modeling, would have an importance on par with that of advancing the technology itself.

5.5 RF MEMS Oscillators

Oscillators in portable wireless systems come in two contexts: stand-alone stable frequency references and voltage-controlled oscillators (VCOs) forming part of a synthesizer phase-locked loop (PLL). Traditionally, the best oscillators, based on the spectral purity measure of phase noise, utilize quartz crystal resonators [40]. Unfortunately, since quartz resonators are not integrable with conventional integrated circuit processes, they must be brought in as external off-chip components. On the other hand, most of the components in a VCO are integrable, except for the varactor. Since off-chip components contribute to a more expensive system, research in the area of developing on-chip RF MEMS–based resonators and varactors has recently received considerable attention [41–46]. In this section we undertake three case studies, namely, that of a micromechanical oscillator, that of a micromachined resonator-based oscillator, and that of an RF MEMS varactor-based VCO.

5.5.1   RF MEMS Oscillator Fundamentals

The properties of MEM resonators and varactors have been discussed extensively in Chapter 3 of this book and in [32]. MEMS-based resonators have two natures: micromechanical and cavity-based. Since the maximum resonance frequency of integrated micromechanical resonators has not exceeded 200 MHz, the oscillator topologies utilized are well known as they pertain to those found in the VHF-UHF range (10–100 MHz) [8, 9] (Figure 5.24), which can exhibit phase noise performance better than −150 dBc/Hz at 1 kHz offset from a 10-MHz carrier [40]. These topologies consist of an amplifier-resonator feedback loop and a buffer amplifier to maintain insensitivity to load variations while delivering the power to the output load. Its design is based on achieving the Barkhausen criterion: oscillation occurs at the frequency at which the loop gain is unity in magnitude and has a phase of 180° [47]. Frequency stability is given by [47]

$$S_f = \frac{d\theta}{d\omega}\bigg|_{\omega_0} \tag{5.39}$$

which for an RLC resonator adopts the value $S_f = -2Q$. Since micromechanical resonators are excited and sensed via capacitive coupling [32] (i.e., via a variable capacitance, C_d), the magnitude of the sense current I_s contains a term that is proportional to the dc voltage V_P across the sense capacitance; that is,

$$I_S = \overline{C}_d \frac{d(\delta V_d)}{dt} + V_P \frac{d(\delta C_d)}{dt} \tag{5.40}$$

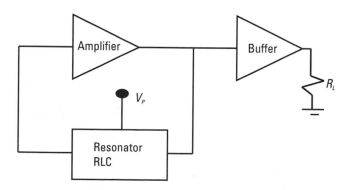

Figure 5.24 Typical oscillator circuit topology for UHF-VHF applications.

Thus, to increase detection sensitivity, it is necessary to include a polarization voltage V_P [32], which is usually on the order of tens of volts. It is important to keep this fact in mind.

On the other hand, given that micromachined cavity resonators are aimed at oscillator applications in the microwave and millimeter-wave frequency range, the oscillator topologies in question are those negative resistance oscillators (Figure 5.25). These oscillators may be described or visualized in terms of a wave bouncing back and forth between a circuit with negative resistance, which possesses a reflection coefficient, $\Gamma_D \gg 1$, and a load, $\Gamma_L < 1$ [Figure 5.25(a)]. In this case the design aims at satisfying a number of conditions. With reference to Figure 5.25(a), the following conditions will be satisfied at the frequency of oscillation [48–51]:

$$R_D(A, \omega) + R_L(\omega) = 0 \qquad (5.41)$$

$$X_D(A, \omega) + X_L(\omega) = 0 \qquad (5.42)$$

where A is the oscillation signal amplitude. Thus, the negative resistance of the active device cancels the dissipation in the load, and the reactance of the active device cancels that of the load—this latter cancellation setting the frequency of oscillation. These conditions are necessary for oscillations to start but are not sufficient to maintain stable oscillations. The following additional condition must be satisfied to ensure stable oscillation [48]:

$$\frac{\partial X_D}{\partial \omega} + \frac{\partial X_L}{\partial \omega} > 0 \qquad (5.43)$$

This condition signifies that the device and load reactances must exhibit a positive slope with respect to the operating point. Figure 5.25(b, c) shows commonly used circuit topologies to implement the negative resistance using a bipolar junction transistor or a field-effect transistor. Essentially, appropriate reactive elements, L, C, or a length of open- or short-circuited transmission line is selected that maximizes the negative resistance, $|R_D| > R_L$, or equivalently, increases the reflection coefficient Γ_D well beyond unity. The resulting oscillator output power capability will be directly proportional to the excess negative resistance beyond the load resistance.

Clearly, since the frequency-determining element in an oscillator is the resonator, varying the resonator's effective inductance or capacitance will change the frequency of oscillation. Most often it is the capacitance that is changed in integrated oscillators; however, when this is done it is at the

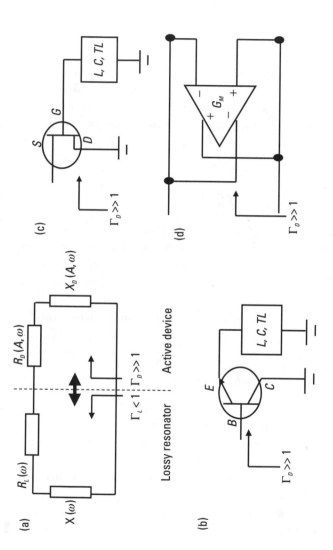

Figure 5.25 (a) Typical oscillator circuit topology for microwave/millimeter-wave applications. Negative resistance implementations: (b) using a bipolar junction transistor; (c) using a field-effect transistor; and (d) using a cross-coupled transconductance amplifier.

expense of increased phase noise due to the usually poor quality factor of on-chip varactors, which ruins the Q of the resonator. The norm, then, has been to utilize off-chip varactors, which then adversely impact assembly cost. Interest in MEMS varactors, therefore, has developed as a promising way to provide a viable alternative for high-Q on-chip varactors [32]. One particularly positive aspect of MEM varactors appears to be that, since they contain no *pn* junctions, there is no danger of forward-biasing them, and phase noise should be improved by mere virtue of allowing a larger oscillation amplitude. This is not a without its limitations, however, since too large a signal amplitude might induce the actuation of the device. The application of a MEM varactor to a negative resistance-type oscillator will be discussed below.

5.5.2 A 14-MHz MEM Oscillator—Case Study

In this section we study the design of a 14-MHz MEM oscillator demonstrated by Mattila et al. [52].

5.5.2.1 Specifications and Topology

The target frequency of the oscillator was set by the resonator. While in principle resonator frequency may be increased indefinitely by reducing its mass [32], thermal energy and contamination limitations prescribe the smallest resonator size to achieve high signal-to-noise ratio [32, 52]. Thus, the resonator utilized in this case consisted of a clamped-clamped mechanical bridge (Figure 5.26) with a bridge length of 44 μm, a bridge width of 4 μm, and coupling capacitive gaps of 0.5 μm, resulting in a resonance frequency of 14 MHz. Accordingly, the oscillator topology of Figure 5.25 was employed.

5.5.2.2 Circuit Design and Implementation

The first step in the circuit design process involved the RF characterization of the resonator. Thus, the resonator was connected in series with an amplifier and network analyzer (Figure 5.27) to measure its transmission characteristics and determine its equivalent circuit. With the MEM resonator placed inside a chamber at a pressure of less than 10^{-2} mbar, the measured resonance lied at $f_0 = 14.3$ MHz and the quality factor was 1,500. This low value of Q was traced to the low length-to-width aspect ratio of the resonator, which was 11 and led to energy leakage through the supporting anchors [52]. The actual oscillator circuit (Figure 5.28) was then obtained by simply removing the network analyzer, narrowbanding the loop with an LC filter of Q~10 center at the resonance frequency, setting the amplifier gain to 33 dB, and closing the loop.

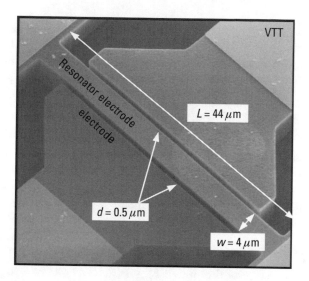

Figure 5.26 A bird's-eye view of the 14-MHz micromechanical bridge resonator. The bridge supports and the outer parts of the electrodes are metallized. (Courtesy of Mr. J. Kiihamäki, VTT Electronics.)

5.5.2.3 Circuit Packaging and Performance

Due to the prototypical nature of the demonstration, the oscillator was put together in a setup rather than packaged (the resonator was inside a chamber). The only effort related to packaging was the careful shielding of the

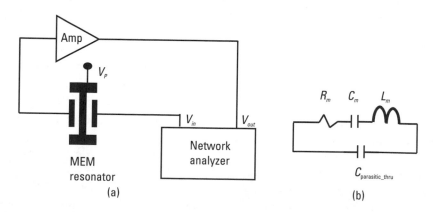

Figure 5.27 (a) Resonator characterization setup; and (b) resonator equivalent circuit: $C_{parasitic\_thru} \sim 17$ fF, $R_m \sim 1$ MΩ $C_m \sim 8$ aF, $L_m \sim 15$H, $C_{gap} \sim 4$ fF. Polarization voltage: $V_P = 100$V. (*After:* [52].)

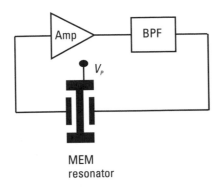

Figure 5.28 14-MHz RF MEM oscillator. (*After:* [52].)

fringing fields to reduce the parasitic feed-through capacitance to ~3.5 fF. The measured performance of the resonator is shown in Figure 5.29. The phase noise at 1 kHz offset is –90 dBc/Hz, and at large frequency offsets it approaches a noise floor of –112 dBc/Hz. From a comparison of the intrinsic resonator noise—set by its equivalent resistance R_m, induced at the amplifier input

$$v^{in}_{n,R_m} = v_{n,R_m} \times \frac{Z(C_{in})}{R_m} = \sqrt{4kTR_m} \times \frac{1/\omega_0 C_{in}}{R_m} \sim 0.4\,nV/\sqrt{Hz},$$

and the measured open-loop gain noise with the resonator switched off ($V_P =$ 0V), $v_n \sim 4.8\,nV/\sqrt{Hz}$—it was determined that the overall oscillator noise was dominated by the amplifier noise. This is indicative of a large noise mismatch and is an aspect that must be addressed to achieve better noise performance. Consideration of these issues led Matilla et al. [52] to propose the following:

1. Increasing the mechanical energy and decreasing the impedance might be accomplished by connecting various resonators in parallel;

2. Optimizing the coupling capacitance C_d by applying a polarization voltage closer to pull-in;

3. Scaling down the amplifier to increase its input impedance and, thus, obtain a higher input impedance that would facilitate noise matching to the resonator.

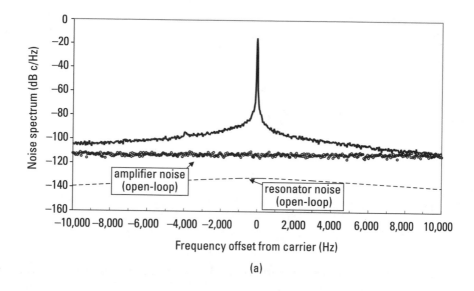

(a)

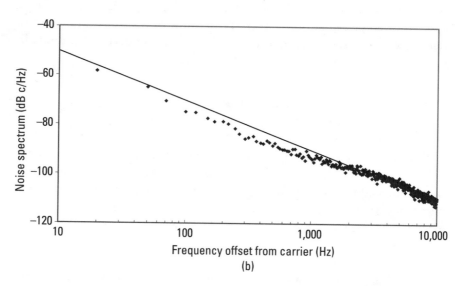

(b)

Figure 5.29 14-MHz RF MEM resonator performance. (a) Output noise spectrum (normalized to carrier, 1-Hz bandwidth) in a ±10-KHz range around the carrier.
The amplifier-induced open-loop noise (measured with the resonator
switched off, $V_P = 0$V) with the same normalization is shown with the open
data points. The simulated intrinsic noise from the resonator dissipation is
shown with the dashed line. (b) The oscillator output noise plotted on a
logarithmic axis. 1/f-dependence of noise decay is indicated using the solid
line. (*Source:* [52] ©2001 Springer-Verlag. Courtesy of Mr. J. Kiihamäki, VTT
Electronics.)

As indicated in [32], contrary to what is commonly observed in crystal oscillators, where the amplitude is set by limiting mechanisms in the active device, in the case of MEM oscillators the amplitude is usually set by non-linearities in the resonator. Thus, Mattila et al. [52] also observed that the oscillation amplitude was limited by the resonator nonlinearities, as evidenced by mechanical spring-hardening. This was partially compensated by the capacitive spring-softening nonlinearity accompanying the applied polarization voltage of $V_P = 100\text{V}$, which had a significant magnitude. An estimate of the mechanical vibration amplitude for a given signal level drive—in this case the signal level at the amplifier input—was obtained as $x_{\text{vib}} \sim v_s C_{in}/\eta \sim 0.05d$, where $\eta = V_P \times \partial C_{\text{gap}}/\partial x$ is the electromechanical coupling constant.

Lessons Learned. (1) At –90 dBc/Hz, the phase noise performance of this 14-MHz MEM resonator-based oscillator is some 60 dB inferior to that of quartz oscillators. This was traced to the reduced mechanical energy concomitant with the small size of the micromechanical resonator, and to difficulty in attaining optimal noise matching due to the characteristically high impedance of the MEM resonators. (2) Either the resonator or the amplifier must be modified so that their respective resistances are closer and noise matching is improved.

5.5.3 A Ka-Band Micromachined Cavity Oscillator—Case Study

In this section we study the design of a 33-GHz micromachined cavity-based oscillator demonstrated by Kwon et al. [53].

5.5.3.1 Specifications and Topology

The Ka-band oscillator was intended for low-cost emerging millimeter-wave commercial applications. In particular, it was directed at solving the problems usually found in dielectric resonator-stabilized oscillators (DROs) and conventional machined cavity-stabilized oscillators, whose usage dominates in these applications. Meeting noise performance specifications is possible with DROs, although at the expense of increased cost due to the high precision with which puck placement must be made, particularly at the higher frequencies [54]. Similarly, the costs incurred in performing conventional high-precision machining to make small cavities are inconsistent with high yield and low-cost manufacturability. By exploiting bulk micromachining techniques, the oscillator topology shown in Figure 5.30 was demonstrated. In terms of the topology introduced earlier, the active device is realized by a

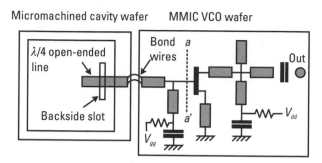

Figure 5.30 Equivalent circuit schematic of micromachined cavity-controlled oscillator. (*Source:* [53] ©1999 IEEE.)

FET-based MMIC, and the lossy resonator is realized by the micromachined cavity.

5.5.3.2 Circuit Design and Implementation

The design of the oscillator derives from common practice as it applies to microwave negative resistance oscillators [48–50]. In this case, we have a common-source 0.2-μm-length gate AlGaAs/InGaAs pseudomorphic HEMT with a total gate periphery of 160 μm, and exhibiting current gain cutoff frequency (f_T) of 70 GHz, and a maximum frequency of oscillation (f_{max}) of 140 GHz, when its drain is biased at 3V. A microstrip stub, in series with the HEMT's source terminal, was used to create a negative resistance at the reference plane delineated by line a-a' at the gate (Figure 5.16). The resonator, in turn, was realized by a micromachined cavity of dimensions 400 μm × 8.4 mm × 5.8 mm, created by wet-etching a double-sided polished (100) silicon wafer in 20% TMAH solution at 90°C. The coupling between the cavity and the active circuit was achieved via a slot defined on the back side of a 250-μm-thick Corning 7740 glass substrate. Essentially, the cavity was capped with the ground plane of the glass wafer, except for a slot in the ground plane, which allowed coupling of the fields in the cavity to an open-circuited $\lambda/4$-long 5-μm-thick gold microstrip stub defined on the top side of the glass substrate. To determine the optimum position and size of the slot for maximum coupling, full-wave simulations were performed. The glass substrate to cavity bonding agent was silver epoxy. Since only three mask levels were utilized, and the major processing steps (namely, deep etching and electroplating) are mature techniques, the process was deemed to be simple and highly manufacturable. Notice, however, that the slot in the back of the metallized glass substrate was etched by ion milling, and that aligning it to the microstrip stub on top required an infrared aligner. The fabricated oscillator is shown in Figure 5.31.

5.5.3.3 Circuit Packaging and Performance

The circuit packaging is illustrated in Figure 5.31. The housing consisted of a metal block containing machined recesses to allow mounting of the resonator and the MMIC at the same level. In addition, a WR-28 waveguide channel was made to guide the oscillator output power. Since the MMIC was implemented using microstrip techniques, coupling to the WR-28 channel was accomplished via a microstrip-to-waveguide transition, which used antipodal fin-lines. The microstrip line that extracts power from the cavity was connected to the MMIC active circuit via a pair of approximately 300-μm-long wirebonds.

Characterization of the cavity resonator yielded an operating frequency of 33.2 GHz, an output power of +10 dBm, a dc-to-RF efficiency of 17%, and a phase noise of −113 dBc/Hz at 1 MHz offset, and −85 dBc/Hz at 100 kHz offset, for a 3.3V drain bias (Figure 5.32).

Quality factor measurement of the cavity yielded a value of just about 130, including losses in the feeding network, but this translates into an

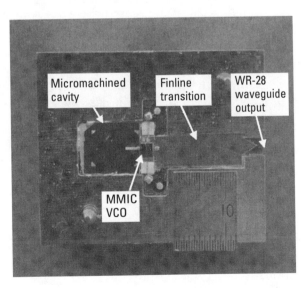

Figure 5.31 Top view of a micromachined cavity oscillator assembled in a WR-28 waveguide package. The scale shown in the photograph is 1 cm. The upper half-block (not shown) is needed to complete the waveguide and the package. (*Source:* [53] ©1999 IEEE.)

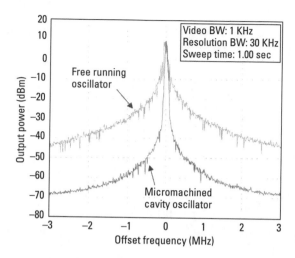

Figure 5.32 Comparison of oscillation spectrum between micromachined cavity oscillator (MCO) and free-running oscillator (FRO) at 2.7V drain bias. The oscillation frequency was 33.2 and 32.8 GHz for the MCO and FRO, respectively.
The bias tuning sensitivity was 20 MHz/V for the MCO and 150 MHz/V for the
FRO. (*Source:* [53] ©1999 IEEE.)

improvement of more than 18 dB in phase noise when compared the performance of a free-running oscillator.

Lessons Learned. (1) Micromachining techniques offer a powerful avenue
to implement low-cost millimeter-wave components, particularly oscillators,
compared to conventional DRO or machined cavity approaches. Their application to micromachined cavities has validated this perception when it
comes to oscillators. (2) The exploitation of micromachining techniques,
however, requires considerable skill in the area of microwave assembly and
package design.

5.5.4   A 2.4-GHz MEMS-Based Voltage-Controlled Oscillator—Case Study

In this section we study the design of a 2.4-GHz MEMS varactor-based
voltage-controlled oscillator (VCO) demonstrated by Dec and Suyama [55].

5.5.4.1   Specifications and Topology

The 2.4-GHz VCO studied in this section was intended for broad application in modern wireless systems [55]. The circuit topology employed is that
of a negative resistance oscillator, where the active device was implemented

by a cross-coupled transconductance amplifier [Figure 5.25(d)], and the *LC* tank resonator was realized with gold wirebonds and the MEM varactors. Figure 5.33 shows the schematic of the VCO circuit.

5.5.4.2 Circuit Design and Implementation

The key circuit components—namely, the cross-coupled transconductance amplifier and the *LC* tank—are shown in Figure 5.34.

The varactor (Figure 5.35) was of the parallel plate type suspended on four springs and fabricated in the MUMPs process [32].

It was designed to achieve the maximum ideal tuning range (i.e., 1.5:1) within a 3.3V tuning voltage. A number of features were implemented to help optimize its yield and performance. For instance, to absorb the residual stress, which tends to warp the top capacitor plate after annealing and sacrificial layer release, curved suspensions fixed to the top plate in the *x-y* plane, which can accommodate its horizontal rotational movement, were utilized. To facilitate gap oxide sacrificial layer release, 141 holes were made on the top plate; holes were also made on the bottom plate to reduce the parasitic capacitance between the bottom plate and the nitride layer. In addition, gold was used wherever possible on the top plate and the suspension beams to maximize the varactor *Q*. The fabricated varactor and its measured performance are shown in Figure 5.36.

The cross-coupled transconductance amplifier was implemented in a 0.5-μm CMOS technology. Its cross-coupled differential pair, formed by transistors M_1 and M_2, realizes the negative transconductance and is biased by the current mirror formed by transistors M_{B1} and M_{B2}. This current mirror, in turn, derives its primary current from an externally supplied reference bias current IBIAS. The design exercise entails choosing the *LC* tank values and the sizes and bias current of the transistors. The *LC* tank is defined by

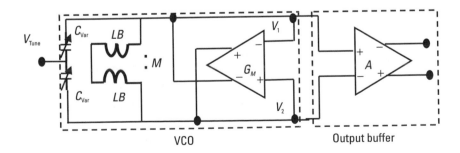

Figure 5.33 MEMS-varactor VCO block diagram. (*After:* [55].)

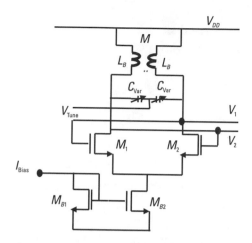

Figure 5.34 MEMS-varactor VCO circuit. (*After:* [55].)

the intended frequency of oscillation, which was arbitrarily chosen to be 2.4 GHz, and the values can be found by optimization, given models for the MEM varactor and circuit parasitics. With the nominal MEM varactor value of 1.4 pF and an estimated parasitic capacitance across the resonant circuit of approximately the same value, the necessary inductance is given by 2.2-nH inductors with a mutual inductance of 0.6 nH. This inductor value is realized by 1-mil-diameter gold wires of 2.2 mm in length and separated by 0.5 mm. The estimated Q of the coupled inductors at 2.4 GHz was 117, while the capacitor Q was 11. The size and bias current of the transistors was chosen as 15.4 mS, seven times the minimum value required for the circuit to initiate oscillation, based on potential process and temperature variation, and 50% varactor Q degradation down.

5.5.4.3 Circuit Packaging and Performance

The assembled VCO is shown in Figure 5.37. The circuit was put together by bonding the MEM varactor and the transconductance amplifier CMOS die onto a ceramic quad flat-pack package.

On the left-hand side of the figure, one can recognize the varactors, which are connected with wirebonds approximately 0.7-mm long and separated by 150 μm to the CMOS die to their right-hand. The coupled inductor wirebond pair is seen to span the CMOS die. The total area occupied by the assembled VCO is 410 × 170 μm. The measured performance of the VCO was as shown in Table 5.4.

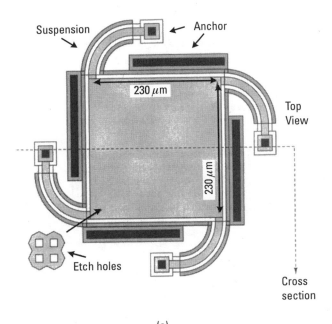

(a)

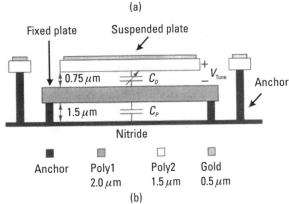

(b)

Figure 5.35 (a) Top and (b) cross-sectional views of the MEM varactor (not shown to scale). The capacitors plates occupy an area of 230 × 230 μm and are separated by an air-gap of 0.75 μm, for a nominal capacitance of 0.6 pF. (*Source:* [55] ©1999 IEEE.)

Environmentally Induced Changes in the Suspended MEM Varactor

One issue that always elicits questions in the context of discussing the application of MEM suspended structures is their susceptibility to external static and dynamic force disturbances. Dec and Suyama [55] conducted a

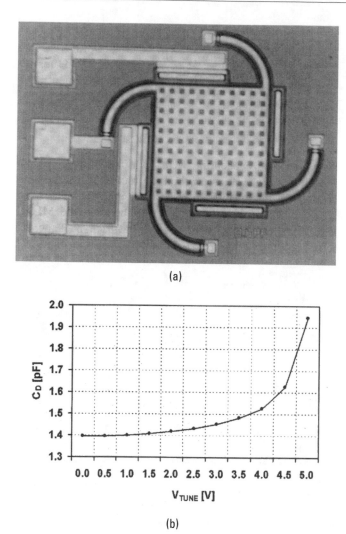

Figure 5.36 (a) Microphotograph and (b) tuning characteristics of MEM varactor. The performance parameters are as follows: nominal capacitance, 1.4 pF; Q-factor at 1 GHz, 23; Q-factor at 2 GHz, 14; tuning range: 1.35:1; tuning voltage, 5V. (*Source:* [55] ©1999 IEEE.)

reasonably detailed analysis on the influence of gravity, acceleration, vibration, and sound on the MEM varactor and the VCO, and their results are gathered here.

As far as the effects of gravity and acceleration, they found, as one would expect, that forces derived from these would, when of the right

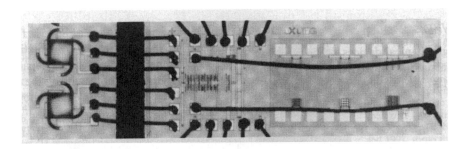

Figure 5.37 Microphotograph of MEM varactor-based VCO. (*Source:* [55] ©1999 IEEE.)

magnitude, cause the top capacitor plate to displace so as to change the air-gap. This deflection, in turn, would cause capacitance and resonance frequency changes given by [55]

$$\frac{\delta C_{\text{Var}}}{C_{\text{Var}}} = -\frac{\delta x}{d + x} \qquad (5.44)$$

and

$$\frac{\delta \omega_0}{\omega_0} = \frac{1}{2} \times \frac{1}{1 + \gamma} \times \frac{\delta x}{d + x} \qquad (5.45)$$

Table 5.4
MEM Varactor-Based VCO Performance

Nominal frequency	2.4 GHz
Tuning range	3.4%
Phase noise at 100 kHz offset	−93 dBc/Hz
Phase noise at 1 MHz offset	−122 dBc/Hz
Output power	−14 dBm
Supply voltage	2.7V
Supply current (VCO core)	5 mA
Supply current (buffer)	15 mA

After: [55].

where γ is the ratio of parasitic to desired capacitance. Application of these formulas to a MEM capacitor like the one utilized in the VCO, which had a mass of $0.8\,\mu$g, an effective stiffness of 39 N/m, and a nominal capacitance of 0.6 pF, yielded the following numerical results. A gravitational acceleration G would cause a displacement $\delta x = mG/k$ equal to 0.2 nm, which, in turn would induced a change of 0.027% in the 0.6-pF capacitance. In addition, the oscillation frequency would register a change of 0.007%, or 170 kHz out of the nominal 2.4 GHz. On the other hand, in order to induce either a 10% change in capacitance or a 2.5% change in resonance frequency, an acceleration of 373g would be necessary. The conclusion from this exercise is that the capacitance and the resonance frequencies may be rather insensitive to reasonable values of accelerations, or that inducing substantial changes in capacitance and frequency would require unreasonably large accelerations.

As far as the influences of vibrations and sound, Dec and Suyama [55] analyzed their respective induced dynamic displacements. The displacement due to an acceleration $G(\omega)$ was expressed as

$$\left|\delta x(\omega)\right| = \left|\frac{X}{F_{\mathrm{exc}}}(\omega)\right| \times m \times \left|G(\omega)\right| \tag{5.46}$$

where $|X(\omega)/F_{\mathrm{exc}}(\omega)|$ is the magnitude of the displacement-to-external-force frequency response, which has a cutoff frequency characterized by the lowest mechanical resonance frequency—in this case, 35 kHz. The resulting displacement, (5.46), modulates the varactor capacitance, which, in turn by (5.45), modulates the VCO frequency of oscillation. Accordingly, a 50-Hz vibration with a peak amplitude of 0.5g would result in a frequency change of approximately 0.0035%, which is a peak deviation of 85 kHz out of the nominal 2.4-GHz oscillation frequency.

The effect of sound on the displacement of the top plate was expressed as

$$\left|\delta x(\omega)\right| = \left|\frac{X}{F_{\mathrm{exc}}}(\omega)\right| \times \left|P_s(\omega)\right| \times \left|G(\omega)\right| \tag{5.47}$$

where $P_s(\omega)$ is the sound pressure impressed upon the plate. Accordingly, a 1-kHz tone 60 dB above the reference 20-mPa sound pressure would induce a displacement of 0.038 nm rms, or equivalently, a change in capacitance of 0.005% rms. This translates, via (5.45), into a frequency deviation of 22 kHz out of the nominal 2.4 GHz. Clearly, this effect could be minimized by, for example, low-pressure packaging.

Lessons Learned. (1) The ingenious layout of multimaterial suspensions, particularly curved suspensions, may allow the dissipation/absorption of in-plane residual stress. (2) Imperviousness to external force disturbances, such as accelerations, vibration, and sound, in addition to resonance frequency and actuation voltage, must be considered when designing the stiffness of suspensions. The problem may be simplified via the decoupling provided by proper packaging (e.g., hermetic packaging reduces sensitivity to sound).

5.6 Summary

In this chapter we have presented a series of case studies. These exercised and integrated the knowledge pertinent to the design of RF MEMS–based circuits gained in previous chapters. The three, perhaps, most important RF MEMS–based circuits were discussed: phase shifters, filters, and oscillators. In particular, for each of these we presented a review of their fundamental topologies and design techniques, followed by a careful step-by-step exposition of the specific approach taken in the realization and implementation of the circuits that were actually fabricated. This entailed discussing their specification and topology, their design and implementation, and their packaging and performance. Lastly, a number of lessons learned were drawn and summarized at the end of each case study, in order to highlight the most important aspects.

References

[1] Terman, F. E., *Radio Engineer's Handbook*, New York: McGraw-Hill, 1943.

[2] Clarke, K. K., and D. T. Hess, *Communication Circuits: Analysis and Design*, Reading, MA: Addison-Wesley, 1971.

[3] Carson, R., *High-Frequency Amplifiers*, New York: John Wiley & Sons, 1975.

[4] Gonzalez, G., *Microwave Transistor Amplifiers, Analysis and Design*, Englewood Cliffs, NJ: Prentice Hall, 1984.

[5] Kraus, J. D., and K. R. Carver, *Electromagnetics*, 2d ed., New York: McGraw-Hill, 1973.

[6] Ramo, S., J. R. Whinnery, and T. Van Duzer, *Fields and Waves on Communication Electronics*, 2d ed., New York: John Wiley & Sons, 1984.

[7] Gupta, K. C., R. Garg, and I. Bahl, *Microstrip Lines and Slot Lines*, 2d ed., Norwood, MA: Artech House, 1996.

[8] Krauss, H. L., C. W. Bostian, and F. H. Raab, *Solid State Radio Engineering*, New York: John Wiley & Sons, 1980.

[9] Hayward, W., *Introduction to Radio Frequency Design*, Englewood Cliffs, NJ: Prentice Hall, 1996.

[10] Collin, R.E., *Foundations of Microwave Engineering*, 2d ed., New York: IEEE Press, 2001.

[11] Matthaei, G. L., Young, L., and E. M. Jones, *Microwave Filters, Impedance-Matching Networks, and Coupling Structures*, Norwood, MA: Artech House, 1980.

[12] Koul, S., and B. Bhat, *Microwave and Millimeter Wave Phase Shifters, Vol. II*, Norwood, MA: Artech House, 1991.

[13] Barker, N. S., and G. M. Rebeiz, "Distributed MEMS True-Time Delay Phase Shifters and Wide-Band Switches," *IEEE Trans. Microwave Theory Tech.*, Vol. 46, November, 1998, pp. 1881–1890.

[14] White, J. F., "High Power P-I-N Diode Controlled Microwave Transmission Line Phase Shifters," *IEEE Trans. Microwave Theory Tech.*, Vol. 13, March 1980, pp. 233–242.

[15] Davis, W. A., *Microwave Semiconductor Circuit Design*, New York: Van Nostrand Reinhold Co., 1984.

[16] Bhat, B., and S. Koul, *Stripline-Like Transmission Lines for Microwave Integrated Circuits*, New Delhi, India: John Wiley & Sons, 1989.

[17] Malczewski, A., et al., "X-Band RF MEMS Phase Shifters for Phased Array Applications," *IEEE Microwave Guided Wave Lett.*, Vol. 9, December 1999, pp. 517–519.

[18] Pillans, B., et al., "Ka-Band RF MEMS Phase Shifters for Phased Array Applications," *2000 IEEE Int. Microwave Symp.*, 2000.

[19] http://www.triquint.com/Mmw/TGP6336/6336.pdf.

[20] Kittel, C., *Introduction to Solid State Physics*, 6th ed., New York: John Wiley & Sons, 1986, p. 144.

[21] Goldsmith, C. L., et al., "Performance of Low-Loss RF MEMS Capacitive Switches," *IEEE Microwave Guided Wave Lett.*, Vol. 8, August 1998, pp. 269–271.

[22] Stenger, O., et al., "A Miniature MMIC One Watt W-Band Solid-State Transmitter," *1997 IEEE Int. Microwave Symp. Digest*, 1997, pp. 431–434.

[23] Kim, M., et al., "A DC-40 GHz Four-Bit RF MEMS True-Time Delay Network," *IEEE Microwave Wireless Components Lett.*, Vol. 11, February 2001, pp. 56–58.

[24] Sokolov, V., et al., A GaAs Monolithic Phase Shifter for 30 GHz Applications," *IEEE MTT Symp.*, 1983, pp. 40–44.

[25] Mihailovich, R. E., et al., "MEM Relay for Reconfigurable RF Circuits," *IEEE Microwave Wireless Components Lett.*, Vol. 11, February 2001, pp. 53–55.

[26] Lakin, K. M., G. R. Kline, and K. T. McCarron, "Development of Miniature Filters for Wireless Applications," *IEEE Trans. Microwave Theory Tech.*, Vol. 43, December 1995, pp. 2933–2939.

[27] Vale, G., et al., "FBAR Filters at GHz Frequencies," *44th Annual Symp. Freq. Control*, 1999, pp. 332–336.

[28] Zverev, A., *Handbook of Filter Synthesis*, New York: John Wiley & Sons, 1967.

[29] Greiner, K., et al., "Integrated RF MEMS for Single Chip Radio," *Transducers'01*, Germany, June 10–14, 2001.

[30] Morkner, H., et al., "An Integrated FBAR Filter and PHEMT Switched-Amp for Wireless Applications," *1999 IEEE Int. Microwave Symp.*, Anaheim, CA, 1999.

[31] Bradley, P., R. Ruby, and J. D. Larson, III, "A Film Bulk Acoustic Resonator (FBAR) Duplexer for USPCS," *2001 IEEE Int. Microwave Symp.*, Phoenix, AZ, 2001.

[32] De Los Santos, H. J., *Introduction to Microelectromechanical (MEM) Microwave Systems*, Norwood, MA: Artech House, 1999.

[33] Brown, A. R., and G. M. Rebeiz, "A High-performance Integrated-Band Diplexer," *IEEE Trans. Microwave Theory. Tech.*, Vol. 47, August 1999, pp. 1477–1481.

[34] Kim, H.-T., et al., "Millimeter-Wave Micromachined Tunable Filters," *1999 IEEE MTT-S Digest*, 1999, pp. 1235–1238.

[35] Bannon, F. D., III, J. R. Clark, and C. T.-C. Nguyen, "High-Q HF Microelectromechanical Filters," *IEEE J. Solid-State Circuits*, Vol. 35, April 2000, pp. 512–526.

[36] Wang, K., and C. T.-C. Nguyen, "High-Order Micromechanical Electronic Filters," *IEEE Micro Electro Mechanical Systems Workshop*, 1999, pp. 25–30.

[37] Nguyen, C. T.-C., and R. T. Howe, "An Integrated CMOS Micromechanical Resonator High-Q Oscillator," *IEEE J. Solid-State Circuits*, Vol. 34, April 1999, pp. 440–445.

[38] Johnson, R. A., *Mechanical Filters in Electronics*, New York, NY: Wiley, 1983.

[39] Konno, M., and H. Nakamura, "Equivalent Electrical Network for the Transversely Vibrating Uniform Bar," *J. Acoust. Soc. Amer.*, Vol. 38, October 1965, pp. 614–622.

[40] Vig, J. R., "Quartz Crystal Resonators and Oscillators for Frequency Control and Timing Applications: A Tutorial," Army Res. Lab, Rep. No. SCLET-TR-88-1, August 1994.

[41] Dec, A., and K. Suyama, "Micromachined Electro-Mechanically Tunable Capacitors and Their Applications to RF," *IEEE Trans. Microwave Theory Tech.*, Vol. 46, December 1998, pp. 2587–2596.

[42] Wang, K., et al., "VHF Free-Free Beam High-*Q* Micromechanical Resonators," *Technical Digest, 12th Int. IEEE Micro Electro Mechanical Syst. Conf.*, Orlando, FL, January 17–21, 1999, pp. 453–458.

[43] Hoivik, N., et al., "Digitally Controllable Variable High-*Q* MEMS Capacitor for RF Applications," *2001 IEEE Int. Microwave Symp.*, Phoenix, AZ, 2001.

[44] Ketterl, T., T. Weller, and D. Fries, "A Micromachined Tunable CPW Resonator," *2001 IEEE Int. Microwave Symp.*, Phoenix, AZ, 2001.

[45] Hsu, W. T., J. R. Clark, and C. T.-C. Nguyen, "A Sub-Micron Capacitive Gap Process for Multiple-Metal-Electrode Lateral Micromechanical Resonators," *2001 IEEE Int. Microwave Symp.*, Phoenix, AZ, 2001.

[46] Lee, S., M. U. Demirci, and. C. T.-C. Nguyen, "A 10-MHz Micromechanical Resonator Pierce Reference Oscillator for Communications," *Transducers'01*, Germany, June 10–14.

[47] Millman, J., and C. C. Halkias, *Integrated Electrinics: Analog and Digital Circuits and Systems*, New York: McGraw-Hill, 1972.

[48] Boyles, J. W., "The Oscillator As a Reflection Amplifier: An Intuitive Approach to Oscillator Design," *Microwave J.*, June 1986, pp. 83–98.

[49] Dougherty, R. M., "MMIC Oscillator Design Techniques," *Microwave J.*, August 1989, pp. 161–162.

[50] "Microwave Oscillator Design," *Hewlett-Packard* Application Note A008.

[51] Kurokawa, K., "Some Basic Characteristics of Broadband Negative Resistance Oscillator Circuits," *Bell Syst. Tech. J.*, Vol. 48, July 1969, pp. 1937–1955.

[52] Mattila, T., et al., "14 MHz Micromechanical Oscillator," *Transducers'01*, Germany, June 10–14, 2001.

[53] Kwon, Y., et al., "A Ka-Band MMIC Oscillator Stabilized with a Micromachined Cavity," *IEEE Microwave Guided Wave Letts.*, Vol. 9, September 1999, pp. 360–362.

[54] Ho, C. Y., and T. Kajita, "DRO State of the Art," *Applied Microwave*, Spring 1990, pp. 69–80.

[55] Dec., A., and K. Suyama, "Microwave MEMS-Based Voltage-Controlled Oscillators," *IEEE Trans. Microwave Theory Tech.*, Vol. 48, November 2000, pp. 1943–1949.

Appendix A: GSM Radio Transmission and Reception Specifications

This appendix is an excerpt of the GSM standard, "Digital cellular telecommunications system (Phase 2+); Radio transmission and reception."

(GSM 05.05 version 8.5.1 Release 1999.) For the complete details of the standards, please refer to the Web site below.

[© ETSI 2000. Further use, modification, or redistribution is strictly prohibited. ETSI standards are available from publication@etsi.fr, and http://www.etsi.org/eds/home.htm.]

Transmitter Characteristics

Throughout this clause, unless otherwise stated, requirements are given in terms of power levels at the antenna connector of the equipment. For equipment with integral antenna only, a reference antenna with 0 dBi gain shall be assumed.

For GMSK modulation, the term output power refers to the measure of the power when averaged over the useful part of the burst (see annex B).

For 8-PSK modulation, the term output power refers to a measure that, with sufficient accuracy, is equivalent to the long term average of the power when taken over the useful part of the burst for random data.

The term peak hold refers to a measurement where the maximum is taken over a sufficient time that the level would not significantly increase if the holding time were longer.

A.1   Output Power

A.1.1   Mobile Station

The MS maximum output power and lowest power control level shall be, according to its class, as defined in the following tables (see also GSM 02.06).

For GMSK Modulation

Power Class	GSM 400 & GSM 900 & GSM 850 Nominal Maximum Output Power	DCS 1800 Nominal Maximum Output Power	PCS 1900 Nominal Maximum Output Power	Tolerance (dB) for Conditions	
1	—	1W (30 dBm)	1W (30 dBm)	±2	±2.5
2	8W (39 dBm)	0.25W (24 dBm)	0.25W (24 dBm)	±2	±2.5
3	5W (37 dBm)	4W (36 dBm)	2W (33 dBm)	±2	±2.5
4	2W (33 dBm)	—	—	±2	±2.5
5	0.8W (29 dBm)	—	—	±2	±2.5

For PSK Modulation

Power Class	GSM 400 & GSM 900 & GSM 850 Nominal Maximum Output Power	GSM 400 & GSM 900 & GSM 850 Tolerance (dB) for Conditions		DCS 1800 Nominal Maximum Output Power	PCS 1900 Nominal Maximum Output Power	DCS 1800 & PCS 1900 Tolerance (dB) for Conditions	
		Normal	Extreme			Normal	Extreme
E1	33 dBm	±2	±2.5	30 dBm	30 dBm	±2	±2.5
E2	27 dBm	±3	±4	26 dBm	26 dBm	−4/+3	±2.5
E3	23 dBm	±3	±4	22 dBm	22 dBm	±3	±4

Maximum output power for 8-PSK in any one band is always equal to or less than GMSK maximum output power for the same equipment in the same band.

A multi band MS has a combination of the power class in each band of operation from the table above. Any combination may be used.

The PCS 1900, including its actual antenna gain, shall not exceed a maximum of 2 Watts (+33 dBm) EIRP per the applicable FCC rules for wideband PCS services [FCC Part 24, Subpart E, Section 24.232]. Power Class 3 is restricted to transportable or vehicular mounted units.

For GSM 850 MS, including its actual antenna gain, shall not exceed a maximum of 7 Watts (+38.5 dBm) ERP per the applicable FCC rules for public mobile services [FCC Part 22, Subpart H, Section 22.913].

The different power control levels needed for adaptive power control (see GSM 05.08) shall have the nominal output power as defined in the table below, starting from the power control level for the lowest nominal output power up to the power control level for the maximum nominal output power corresponding to the class of the particular MS as defined in the table above. Whenever a power control level commands the MS to use a nominal output power equal to or greater than the maximum nominal output power for the power class of the MS, the nominal output power transmitted shall be the maximum nominal output power for the MS class, and the tolerance specified for that class (see table above) shall apply.

GSM 400 and GSM 900 and GSM 850

Power Control Level	Nominal Output Power (dBm)	Tolerance (dB) for Conditions	
		Normal	Extreme
0–2	39	±2	±2.5
3	37	±3	±4
4	36	±3	±4
5	33	±3	±4
6	32	±3	±4
7	29	±3	±4
8	27	±3	±4
9	25	±3	±4
10	23	±3	±4

GSM 400 and GSM 900 and GSM 850 (continued)

Power Control Level	Nominal Output Power (dBm)	Tolerance (dB) for Conditions	
		Normal	Extreme
11	21	±3	±4
12	19	±3	±4
13	17	±3	±4
14	15	±3	±4
15	13	±3	±4
16	11	±5	±6
17	9	±5	±6
18	7	±5	±6
19–31	5	±5	±6

DCS 1800

Power Control Level	Nominal Output Power (dBm)	Tolerance (dB) for Conditions	
		Normal	Extreme
29	36	±2	±2.5
30	34	±3	±4
31	32	±3	±4
0	30	±3	±4
1	28	±3	±4
2	26	±3	±4
3	24	±3	±4
4	22	±3	±4
5	20	±3	±4
6	18	±3	±4
7	16	±3	±4
8	14	±3	±4

DCS 1800 (continued)

Power Control Level	Nominal Output Power (dBm)	Tolerance (dB) for Conditions	
		Normal	Extreme
9	12	±4	±5
10	10	±4	±5
11	8	±4	±5
12	6	±4	±5
13	4	±4	±5
14	2	±5	±6
15–28	0	±5	±6

NOTE 1: For DCS 1800, the power control levels 29, 30, and 31 are not used when transmitting the parameter MS_TXPWR_MAX_CCH on BCCH, for cross phase compatibility reasons. If levels greater than 30 dBm are required from the MS during a random access attempt, then these shall be decoded from parameters broadcast on the BCCH as described in GSM 05.08.

Furthermore, the difference in output power actually transmitted by the MS between two power control levels where the difference in nominal output power indicates an increase of 2 dB (taking into account the restrictions due to power class), shall be +2 ±1.5 dB. Similarly, if the difference in output power actually transmitted by the MS between two power control levels where the difference in nominal output power indicates an decrease of 2 dB (taking into account the restrictions due to power class), shall be −2 ±1.5 dB.

NOTE 2: A 2-dB nominal difference in output power can exist for non-adjacent power control levels, e.g., power control levels 18 and 22 for GSM 400 and GSM 900; power control levels 31 and 0 for class 3 DCS 1800 and power control levels 3 and 6 for class 4 GSM 400 and GSM 900.

A change from any power control level to any power control level may be required by the base transmitter. The maximum time to execute this change is specified in GSM 05.08.

PCS 1900

Power Control Level	Nominal Output Power (dBm)	Tolerance (dB) for Conditions	
		Normal	Extreme
22–29	Reserved	Reserved	Reserved
30	33	±2 dB	±2.5 dB
31	32	±3 dB1	±2.5 dB
0	30	±3 dB	±4 dB1
1	28	±3 dB	±4 dB
2	26	±3 dB1	±4 dB
3	24	±3 dB	±4 dB1
4	22	±3 dB	±4 dB
5	20	±3 dB	±4 dB
6	18	±3 dB	±4 dB
7	16	±3 dB	±4 dB
8	14	±3 dB	±4 dB
9	12	±4 dB	±5 dB
10	10	±4 dB	±5 dB
11	8	±4 dB	±5 dB
12	6	±4 dB	±5 dB
13	4	±4 dB	±5 dB
14	2	±5 dB	±6 dB
15	0	±5 dB	±6 dB
16–21	Reserved	Reserved	Reserved

NOTE: Tolerance for MS Classes 1 and 2 is ±2 dB normal and ±2.5 dB extreme at Power Control Levels 0 and 3, respectively.

The output power actually transmitted by the MS at each of the power control levels shall form a monotonic sequence, and the interval between power steps shall be 2 dB ±1.5 dB except for the step between power control levels 30 and 31 where the interval is 1 dB ±1 dB.

The MS transmitter may be commanded by the BTS to change from any power control level to any other power control level. The maximum time to execute this change is specified in GSM 05.08.

For CTS transmission, the nominal maximum output power of the MS shall be restricted to:

- 11 dBm (0.015 W) in GSM 900, i.e., power control level 16;
- 12 dBm (0.016 W) in DCS 1800, i.e., power control level 9.

A.1.2 Base Station

The Base Station Transmitter maximum output power, at GMSK modulation, measured at the input of the BSS Tx combiner, shall be, according to its class, as defined in the following tables.

GSM 400 & GSM 900 & GSM 850 & MXM 850

TRX Power Class	Maximum Output Power
1	320–(<640)W
2	160–(<320)W
3	80–(<)W
4	40–(<80)W
5	20–(<40)W
6	10–(<20)W
7	5–(<10)W
8	2.5–(<5)W

DCS 1800 & PCS 1900 & MXM 1900

TRX Power Class	Maximum Output Power
1	20–(<40)W
2	10–(<20)W
3	5–(<10)W
4	12.5–(<5)W

The micro-BTS maximum output power per carrier measured at the antenna connector after all stages of combining shall be, according to its class, defined in the following table.

GSM 900 & GSM 850 & MXM 850 Micro- and Pico-BTS		DCS 1800 & PCS 1900 & MXM 1900 Micro- and Pico-BTS	
TRX power class	Maximum output power	TRX power class	Maximum output power
Micro	—	Micro	—
M1	(>19)–24 dBm	M1	(>27)–32 dBm
M2	(>14)–19 dBm	M2	(>22)–27 dBm
M3	(>9)–14 dBm	M3	(>17)–22 dBm
Pico P1	(>13)–20 dBm	Pico P1	(>16)–23 dBm

For BTS supporting 8-PSK, the manufacturer shall declare the output power capability at 8-PSK modulation. The class of a micro-BTS or a pico-BTS is defined by the highest output power capability for either modulation and the output power shall not exceed the maximum output power of the corresponding class.

The tolerance of the actual maximum output power of the BTS shall be ±2 dB under normal conditions and ±2.5 dB under extreme conditions. Settings shall be provided to allow the output power to be reduced from its maximum level in at least six steps of nominally 2 dB with an accuracy of ±1 dB to allow a fine adjustment of the coverage by the network operator. In addition, the actual absolute output power at each static RF power step (N) shall be $2 \times N$ dB below the absolute output power at static RF power step 0 with a tolerance of ±3 dB under normal conditions and ±4 dB under extreme conditions. The static RF power step 0 shall be the actual output power according to the TRX power class.

As an option the BSS can utilize downlink RF power control. In addition to the static RF power steps described above, the BSS may then utilize up to 15 steps of power control levels with a step size of 2 dB ±1.5 dB, in addition the actual absolute output power at each power control level (N) shall be $2 \times N$ dB below the absolute output power at power control level 0 with a tolerance of ±3 dB under normal conditions and ±4 dB under extreme conditions. The power control level 0 shall be the set output power according to the TRX power class and the six power settings defined above. Network operators or manufacturers may also specify the BTS output power including any Tx combiner, according to their needs.

A.1.2.1 Additional Requirements for PCS 1900 and MXM 1900 Base Stations

The BTS transmitter maximum rated output power per carrier, measured at the input of the transmitter combiner, shall be, according to its TRX power class, as defined in the table above. The base station output power may also be specified by the manufacturer or system operator at a different reference point (e.g. after transmitter combining).

The maximum radiated power from the BTS, including its antenna system, shall not exceed a maximum of 1640 W EIRP, equivalent to 1000 W ERP, per the applicable FCC rules for wideband PCS services [FCC part 24, subpart E, clause 24.237].

A.1.2.2 Additional Requirements for GSM 850 and MXM 850 Base Stations

The BTS transmitter maximum rated output power per carrier, measured at the input of the transmitter combiner, shall be, according to its TRX power class, as defined in the table above. The base station output power may also be specified by the manufacturer or system operator at a different reference point (e.g., after transmitter combining). The maximum radiated power from the BTS, including its antenna system, shall not exceed a maximum of 500 W ERP, per the applicable FCC rules for public mobile services [FCC part 22, subpart H, clause 22.913].

A.2 Output RF Spectrum

The specifications contained in this clause apply to both BTS and MS, in frequency hopping as well as in non frequency hopping mode, except that beyond 1800 kHz offset from the carrier the BTS is not tested in frequency hopping mode. Due to the bursty nature of the signal, the output RF spectrum results from two effects:

- The modulation process;
- The power ramping up and down (switching transients).

The two effects are specified separately; the measurement method used to analyze separately those two effects is specified in GSM 11.10 and 11.21. It is based on the "ringing effect" during the transients, and is a measurement in the time domain, at each point in frequency.

The limits specified thereunder are based on a 5-pole synchronously tuned measurement filter.

Unless otherwise stated, for the BTS, only one transmitter is active for the tests of this clause.

A.2.1   Spectrum Due to Modulation and Wideband Noise

The output RF modulation spectrum is specified in the following tables. A mask representation of the present document is shown in annex A. The present document applies for all RF channels supported by the equipment.

The specification applies to the entire of the relevant transmit band and up to 2 MHz either side.

The specification shall be met under the following measurement conditions:

- For BTS up to 1800 kHz from the carrier and for MS in all cases:

- Zero frequency scan, filter bandwidth and video bandwidth of 30 kHz up to 1800 kHz from the carrier and 100 kHz at 1800 kHz and above from the carrier, with averaging done over 50% to 90% of the useful part of the transmitted bursts, excluding the midamble, and then averaged over at least 200 such burst measurements. Above 1800 kHz from the carrier only measurements centred on 200 kHz multiples are taken with averaging over 50 bursts.

- For BTS at 1800 kHz and above from the carrier:

- Swept measurement with filter and video bandwidth of 100 kHz, minimum sweep time of 75 ms, averaging over 200 sweeps. All slots active, frequency hopping disabled.

- When tests are done in frequency hopping mode, the averaging shall include only bursts transmitted when the hopping carrier corresponds to the nominal carrier of the measurement. The specifications then apply to the measurement results for any of the hopping frequencies.

The figures in tables a) and b) below, at the vertically listed power level (dBm) and at the horizontally listed frequency offset from the carrier (kHz), are then the maximum allowed level (dB) relative to a measurement in 30 kHz on the carrier.

NOTE: This approach of specification has been chosen for convenience and speed of testing. It does, however, require careful interpretation if there is a need to convert figures in the following tables into spectral density values, in that only part of the power of the carrier is used as the relative reference, and in addition different measurement bandwidths are applied at different offsets from the carrier. Appropriate conversion factors for this purpose are given in GSM 05.50.

For the BTS, the power level is the "actual absolute output power" defined in clause 4.1.2. If the power level falls between two of the values in the table, the requirement shall be determined by linear interpolation.

a1) GSM 400 and GSM 900 and GSM 850 MS:

	100	200	250	400	= 600 < 1800	=1800 < 3000	= 3000 < 6000	= 6000
= 39	+0.5	−30	−33	−60	−66	−69	−71	−77
37	+0.5	−30	−33	−60	−64	−67	−69	−75
36	+0.5	−30	−33	−60	−62	−65	−67	−73
= 33	+0.5	−30	−33	−60*	−60	−63	−65	−71

NOTE: *For equipment supporting 8-PSK, the requirement for 8-PSK modulation is −54 dB.

a2) GSM 400 and GSM 900 and GSM 850 and MXM 850 Normal BTS:

	100	200	250	400	=600 < 1200	=1200 < 1800	=1800 < 6000	= 6000
= 43	+0.5	−30	−33	−60*	−70	−73	−75	−80
41	+0.5	−30	−33	−60*	−68	−71	−73	−80
39	+0.5	−30	−33	−60*	−66	−69	−71	−80
37	+0.5	−30	−33	−60*	−64	−67	−69	−80
35	+0.5	−30	−33	−60*	−62	−66	−67	−80
= 33	+0.5	−30	−33	−60*	−60	−63	−65	−80

NOTE: *For equipment supporting 8-PSK, the requirement for 8-PSK modulation is −56 dB.

a3) GSM 900 and GSM 850 and MXM 850 Micro-BTS:

	100	200	250	400	= 600 < 1200	=1200 < 1800	=1800
= 33	+0.5	−30	−33	−60*	−60	−63	−70

NOTE: * For equipment supporting 8-PSK, the requirement for 8-PSK modulation is −56 dB.

a4) GSM 900 and GSM 850 and MXM 850 Pico-BTS:

	100	200	250	400	600 ≥ < 1200	1200 ≥ < 1800	1800 ≥ < 6000	≥ 6000
= 20	+0.5	−30	−33	−60*	−60	−63	−70	−80

NOTE: * For equipment supporting 8-PSK, the requirement for 8-PSK modulation is −56 dB.

b1) DCS 1800 MS:

	100	200	250	400	≥ 600 < 1800	≥ 1800 < 6000	≥ 6000
≥ 36	+0.5	−30	−33	−60	−60	−71	−79
34	+0.5	−30	−33	−60	−60	−69	−77
32	+0.5	−30	−33	−60	−60	−67	−75
30	+0.5	−30	−33	−60*	−60	−65	−73
28	+0.5	−30	−33	−60*	−60	−63	−71
26	+0.5	−30	−33	−60*	−60	−61	−69
≥ 24	+0.5	−30	−33	−60*	−60	−59	−67

NOTE: *For equipment supporting 8-PSK, the requirement for 8-PSK modulation is −54 dB.

b2) DCS 1800 Normal BTS:

	100	200	250	400	≥ 600 < 1200	≥1200 < 1800	≥ 1800 < 600	≥ 6000
≥ 43	+0.5	−30	−33	−60*	−70	−73	−75	−80
41	+0.5	−30	−33	−60*	−68	−71	−73	−80
39	+0.5	−30	−33	−60*	−66	−69	−71	−80
37	+0.5	−30	−33	−60*	−64	−67	−69	−80
35	+0.5	−30	−33	−60*	−62	−65	−67	−80
≥ 33	+0.5	−30	−33	−60*	−60	−63	−65	−80

NOTE: *For equipment supporting 8-PSK, the requirement for 8-PSK modulation is −56 dB.

b3) DCS 1800 Micro-BTS:

	100	200	250	400	≥ 600 < 1200	≥ 1200 < 1800	≥ 1800
35 ≥ 33	+0.5	−30	−33	−60*	−628	−65	−76
	+0.5	−30	−33	−60*	−60	−63	−76

NOTE: *For equipment supporting 8-PSK, the requirement for 8-PSK modulation is −56 dB.

b4) DCS 1800 Pico-BTS:

	100	200	250	400	≥ 600 < 1200	≥ 1200 < 1800	≥ 1800 < 6000	≥ 6000
≥ 23	+0.5	−30	−33	−60*	−60	−63	−76	−80

NOTE: *For equipment supporting 8-PSK, the requirement for 8-PSK modulation is −56 dB.

c1) PCS 1900 MS:

	100	200	250	400	≥ 600 < 1200	≥ 1200 < 1800	≥ 1800 < 6000	≥ 6000
≥ 33	+0.5	−30	−33	−60	−60	−60	−68	−76
32	+0.5	−30	−33	−60	−60	−60	−67	−75
30	+0.5	−30	−33	−60*	−60	−60	−66	−73
28	+0.5	−30	−33	−60*	−60	−60	−63	−71
26	+0.5	−30	−33	−60*	−60	−60	−61	−69
≥ 24	+0.5	−30	−33	−60*	−60	−60	−59	−67

NOTE: *For equipment supporting 8-PSK, the requirement for 8-PSK modulation is −54 dB.

c2) PCS 1900 & MXM 1900 Normal BTS:

	100	200	250	400	≥ 600 < 1200	≥ 120 < 1800	≥ 1800 < 6000	≥ 6000
≥ 43	+0.5	−30	−33	−60*	−70	−73	−75	−80
	+0.5	−30	−33	−60*	−68	−71	−73	−80
	+0.5	−30	−33	−60*	−66	−69	−71	−80
	+0.5	−30	−33	−60*	−64	−67	−69	−80
	+0.5	−30	−33	−60*	−62	−65	−67	−80
	+0.5	−30	−33	−60*	−60	−63	−66	−80

NOTE: *For equipment supporting 8-PSK, the requirement for 8-PSK modulation is −56 dB.

c3) PCS 1900 & MXM 1900 Micro-BTS:

	100	200	250	400	≥ 600 < 1200	≥ 1200 < 1800	≥ 1800
35	+0.5	−30	−33	−60*	−62	−65	−76
≥ 33	+0.5	−30	−33	−60*	−60	−63	−76

NOTE: *For equipment supporting 8-PSK, the requirement for 8-PSK modulation is −56 dB.

The PCS 1900 micro-BTS spectrum due to modulation and noise at all frequency offsets greater than 1.8 MHz from carrier shall be −76 dB for all micro-BTS classes. These are average levels in a measurement bandwidth of 100 kHz relative to a measurement in 30 kHz on carrier. The measurement will be made in nonfrequency hopping mode under the conditions specified for the normal BTS.

c4) PCS 1900 and MXM 1900 Pico-BTS:

	100	200	250	400	≥ 600 < 1200	≥ 1200 < 1800	≥ 1800
≥ 23	+0.5	−30	−33	−60*	−60	−63	−76

NOTE: *For equipment supporting 8-PSK, the requirement for 8-PSK modulation is −56 dB.

The following exceptions shall apply, using the same measurement conditions as specified above.

i. In the combined range 600 kHz to 6 MHz above and below the carrier, in up to three bands of 200 kHz width centered on a frequency which is an integer multiple of 200 kHz, exceptions at up to −36 dBm are allowed.

ii. Above 6 MHz offset from the carrier in up to 12 bands of 200 kHz width centered on a frequency which is an integer multiple of 200 kHz, exceptions at up to −36 dBm are allowed. For the BTS only one transmitter is active for this test.

Using the same measurement conditions as specified above, if a requirement in tables a*x*), b*x*), and c*x*) is tighter than the limit given in the following, the latter shall be applied instead.

iii. For MS:

Frequency Offset from the Carrier	GSM 400 & GSM 900 & GSM 850	DCS 1800 & PCS 1900
< 600 kHz	−36 dBm	−36 dBm
≥ 600 kHz, < 1800 kHz	−51 dBm	−56 dBm
≥ 1800 kHz	−46 dBm	−51 dBm

iv. For normal BTS, whereby the levels given here in dB are relative to the output power of the BTS at the lowest static power level measured in 30 kHz:

Frequency Offset from the Carrier	GSM 400 & GSM 900 & GSM 850 & MXM 850	DCS 1800 & PCS 1900 & MXM 1900
< 1800 kHz	Max{−85 dB, −65 dBm}	Max{−88 dB, −57 dBm}
≥ 1800 kHz	Max{−83 dB, −65 dBm}	Max{−83 dB, −57 dBm}

v. For micro- and pico-BTS, at 1800 kHz and above from the carrier:

Power Class	GSM 900 & GSM 850 & MXM 850	DCS 1800 & PCS 1900 & MXM 1900
M1	−59 dBm	−57 dBm
M2	−64 dBm	−62 dBm
M3	−69 dBm	−67 dBm
P1	−68 dBm	−65 dBm

A.2.2 Spectrum Due to Switching Transients

Those effects are also measured in the time domain and the specifications assume the following measurement conditions: zero frequency scan, filter bandwidth 30 kHz, peak hold, and video bandwidth 100 kHz. The example of a waveform due to a burst as seen in a 30 kHz filter offset from the carrier is given there under (Figure A.1).

a) Mobile Station:

Power Level	Maximum Level Measured			
	400 kHz	600 kHz	1200 kHz	1800 kHz
39 dBm	−21 dBm	−26 dBm	−32 dBm	−36 dBm
≥ 37 dBm	−23 dBm	−26 dBm	−32 dBm	−36 dBm

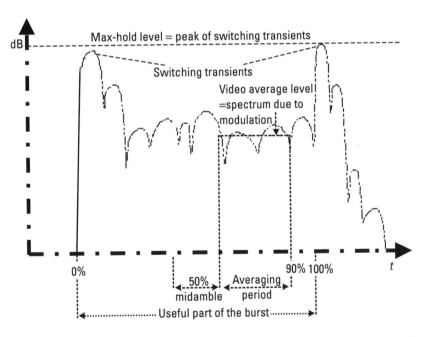

Figure A.1 Example of a time waveform due to a burst as seen in a 30-kHz filter offset from the carrier.

NOTE 1: The relaxation for power level 39 dBm is in line with the modulated spectra and thus causes negligible additional interference to an analogue system by a GSM signal.

NOTE 2: The near-far dynamics with the present document has been estimated to be approximately 58 dB for MS operating at a power level of 8W or 49 dB for MS operating at a power level of 1W. The near-far dynamics then gradually decreases by 2 dB per power level down to 32 dB for MS operating in cells with a maximum allowed output power of 20 mW or 29 dB for MS operating at 10 mW.

NOTE 3: The possible performance degradation due to switching transient leaking into the beginning or the end of a burst was estimated and found to be acceptable with respect to the BER due to cochannel interference (C/I).

b) Base transceiver station: The maximum level measured, after any filters and combiners, at the indicated offset from the carrier, is:

	Maximum Level Measured			
	400 kHz	600 kHz	1200 kHz	1800 kHz
GSM 400 & GSM 900 & GSM 850 & MXM 850 (GMSK)	−67 dBc	−67 dBc	−74 dBc	−74 dBc
GSM 400 & GSM 900 & GSM 850 & MXM 850 (8-PSK)	−62 dBc	−62 dBc	−74 dBc	−74 dBc
DCS 1800 & PCS 1900 & MXM 1900 (GMSK)	−50 dBc	−58 dBc	−66 dBc	−66 dBc
DCS 1800 & PCS 1900 & MXM 1900 (8-PSK)	−60 dBc	−58 dBc	−66 dBc	−66 dBc

Or −36 dBm, whichever is the higher.

dBc means relative to the output power at the BTS, measured at the same point and in a filter bandwidth of at least 300 kHz.

NOTE 4: Some of the above requirements are different from those specified in clause A.3.2.

A.3 Spurious Emissions

The limits specified thereunder are based on a 5-pole synchronously tuned measurement filter.

In addition to the requirements of this clause, the PCS 1900 & MXM 1900 BTS and PCS 1900 MS shall also comply with the applicable limits for spurious emissions established by the FCC rules for wideband PCS services. In addition to the requirements of this clause, the GSM 850 & MXM 850 BTS and GSM 850 MS shall also comply with the applicable limits for spurious emissions established by the FCC rules for public mobile services [FCC Part 22, Subpart H].

A.3.1 Principle of the Specification

In this clause, the spurious transmissions (whether modulated or unmodulated) and the switching transients are specified together by measuring the peak power in a given bandwidth at various frequencies. The bandwidth is increased as the frequency offset between the measurement frequency, and either the carrier, or the edge of the MS or BTS transmit band increases. The effect for spurious signals of widening the measurement bandwidth is to reduce the allowed total spurious energy per MHz. The effect for switching transients is to effectively reduce the allowed level of the switching transients (the peak level of a switching transient increases by 6 dB for each doubling of the measurement bandwidth). The conditions are specified in the following table, a peak-hold measurement being assumed.

The measurement conditions for radiated and conducted spurious are specified separately in GSM 11.10 and 11.2x series. The frequency bands where these are actually measured may differ from one type to the other (see GSM 11.10 and 11.2x series).

a)

Band	Frequency Offset	Measurement Bandwidth
Relevant transmit band	(offset from carrier) ≥ 1.8 MHz	30 kHz
	≥ 6 MHz	100 kHz

b)

Band	Frequency Offset	Measurement Bandwidth
100 kHz to 50 MHz	—	10 kHz
50 MHz to 500 MHz outside the relevant transmit band	(offset from edge of the relevant transmit band)	
	≥ 2 MHz	10 kHz
	≥ 5 MHz	100 kHz
Above 500 MHz outside the relevant transmit band	(offset from edge of the relevant transmit band)	
	≥ 2 MHz	30 kHz
	≥ 5 MHz	100 kHz
	≥ 10 MHz	300 kHz
	≥ 20 MHz	1 MHz
	≥ 30 MHz	3 MHz

The measurement settings assumed correspond, for the resolution bandwidth, to the value of the measurement bandwidth in the table, and for the video bandwidth, to approximately three times this value.

NOTE: For radiated spurious emissions for MS with antenna connectors, and for all spurious emissions for MS with integral antennas, the specifications currently only apply to the frequency band 30 MHz to 4 GHz. The specification and method of measurement outside this band are under consideration.

A.3.2　Base Transceiver Station

A.3.2.1　General Requirements

The power measured in the conditions specified in clause A.3.1a shall be no more than −36 dBm.

The power measured in the conditions specified in clause A.3.1b shall be no more than:

- 250 nW (−36 dBm) in the frequency band 9 kHz to 1 GHz;

- 1 μW (−30 dBm) in the frequency band 1 GHz to 12.75 GHz.

NOTE 1: For radiated spurious emissions for BTS, the specifications currently only apply to the frequency band 30 MHz to 4 GHz. The specification and method of measurement outside this band are under consideration.

In the BTS receive band, the power measured using the conditions specified in clause A.2.1, with a filter and video bandwidth of 100 kHz shall be no more than:

	GSM 900 & GSM 850 & HSM 850 (dBm)	DCS 1800 & PCS 1900 & MXM 1900 (dBm)
Normal BTS	−98	−98
Micro BTS M1	−91	−96
Micro BTS M2	−86	−91
Micro BTS M3	−81	−86
Pico BTS P1	−70	−80
R-GSM 900 BTS	−89	—

These values assume a 30 dB coupling loss between transmitter and receiver. If BTSs of different classes are cosited, the coupling loss must be increased by the difference between the corresponding values from the table above.

Measures must be taken for mutual protection of receivers when BTS of different bands are cosited.

NOTE 2: Thus, for this case, assuming the coupling losses are as above, then the power measured in the conditions specified in clause A.2.1, with a filter and video bandwidth of 100 kHz should be no more than the values in the table above for the GSM 400 and GSM 900 transmitter in the band 1710 MHz to 1785 MHz, for GSM 400 and DCS 1800 transmitter in the band 876 MHz to 915 MHz and for GSM 900 and DCS 1800 transmitter in the bands 450.4 MHz to 457.6 MHz and 478.8 MHz to 486.0 MHz.

In any case, the powers measured in the conditions specified in clause A.2.1, with a filter and video bandwidth of 100 kHz shall be no more than −47 dBm for the GSM 400 and GSM 900 BTS in the band 1805 MHz to

1880 MHz and −57 dBm for a GSM 400 and DCS 1800 BTS in the band 921 MHz to 960 MHz.

Measures must be taken for mutual protection of receivers when MXM 850 and MXM 1900 BTS, or GSM 850 and PCS 1900 BTS are cosited.

NOTE 3: Thus, for this case, assuming the coupling losses are as above, then the power measured in the conditions specified in clause A.2.1, with a filter and video bandwidth of 100 kHz should be no more than the values in the table above for the MXM 850 (or GSM 850 BTS) transmitter in the band 1850 MHz to 1910 MHz and for MXM 1900 (or PCS 1900 BTS) transmitter in the band 824 MHz to 849 MHz.

In any case, the powers measured in the conditions specified in clause A.2.1, with a filter and video bandwidth of 100 kHz shall be no more than −47 dBm for an MXM 850 BTS (or GSM 850 BTS) in the band 1930 MHz to 1990 MHz and −57 dBm for an MXM 1900 BTS (or PCS 1900 BTS) in the band 869 MHz to 894 MHz.

NOTE 4: In addition, to protect cocoverage systems, the powers measured in the conditions specified in clause A.2.1, with a filter and video bandwidth of 100 kHz should be no more than −57 dBm for the GSM 900 and DCS 1800 BTS in the band 460.4 MHz to 467.6 MHz and 488.8 MHz to 496.0 MHz.

A.3.2.2   Additional Requirements for Coexistence with 3G

In geographic areas where GSM and UTRA networks are deployed, the power measured in the conditions specified in clause A.2.1, with a filter and videobandwidth of 100 kHz shall be no more than:

Band (MHz)	Power (dBm)	Note
1900–1920	−62	UTRA/TDD band
1920–1980	−62	UTRA/FDD BS Rx band
2010–2025	−62	UTRA/TDD band
2110–2170	−62	UTRA/FDD UE Rx band

When GSM and UTRA BS are colocated, the power measured in the conditions specified in clause 4.2.1, with a filter and videobandwidth of 100 kHz shall be no more than:

Band (MHz)	Power (dBm)	Note
1900–1920	−96	UTRA/TDD band
1920–1980	−96	UTRA/FDD BS Rx band
2010–2025	−96	UTRA/TDD band
2110–2170	−96	UTRA/FDD UE Rx band

NOTE: The requirements in this clause should also be applied to BTS built to a hardware specification for R98 or earlier. For a BTS built to a hardware specification for R98 or earlier, with an 8-PSK capable transceiver installed, the 8-PSK transceiver shall meet the R99 requirement.

A.3.3 Mobile Station

A.3.3.1 Mobile Station GSM 400, GSM 900, and DCS 1800

The power measured in the conditions specified in clause A.3.1a, for a MS when allocated a channel, shall be no more than −36 dBm. For R-GSM 900 MS except small MS the corresponding limit shall be −42 dBm.

The power measured in the conditions specified in clause A.3.1b for a MS, when allocated a channel, shall be no more than (see also note in clause A.3.1b above):

- 250 nW (−36 dBm) in the frequency band 9 kHz to 1 GHz;
- 1 μW (−30 dBm) in the frequency band 1 GHz to 12.75 GHz.

The power measured in a 100 kHz bandwidth for an MS, when not allocated a channel (idle mode), shall be no more than (see also note in clause A.3.1 above):

- 2 nW (−57 dBm) in the frequency bands 9 kHz to 1000 MHz;
- 20 nW (−47 dBm) in the frequency bands 1–12.75 GHz, with the following exceptions:
- 1.25 nW (−59 dBm) in the frequency band 880 MHz to 915 MHz;
- 5 nW (−53 dBm) in the frequency band 1.71 GHz to 1.785 GHz;
- −76 dBm in the frequency bands 1900–1920 MHz, 1920–1980 MHz, 2010–2025 MHz, and 2210–2170 MHz.

NOTE: The idle mode spurious emissions in the receive band are covered by the case for MS allocated a channel (see below).

When allocated a channel, the power emitted by the MS, when measured using the measurement conditions specified in clause A.2.1, but with averaging over at least 50 burst measurements, with a filter and video bandwidth of 100 kHz, for measurements centred on 200 kHz multiples shall be no more than:

- −67 dBm in the bands 460.4–467.6 MHz and 488.8–496 MHz for GSM400 MS only;

- −60 dBm in the band 921–925 MHz for R-GSM MS only;

- −67 dBm in the band 925–935 MHz;

- −71 dBm in the band 1805–1880 MHz;

- −66 dBm in the bands 1900–1920 MHz, 1920–1980 MHz, 2010–2025 MHz, and 2110–2170 MHz.

As exceptions up to five measurements with a level up to −36 dBm are permitted in each of the bands 925 MHz to 960 MHz, 1805 MHz to 1880 MHz, 1900–1920 MHz, 1920–1980 MHz, 2010–2025 MHz, and 2110–2170 MHz for each ARFCN used in the measurements. For GSM 400 MS, in addition, exceptions up to three measurements with a level up to −36 dBm are permitted in each of the bands 460.4 MHz to 467.6 MHz and 488.8 MHz to 496 MHz for each ARFCN used in the measurements.

When hopping, this applies to each set of measurements, grouped by the hopping frequencies as described in clause A.2.1.

A.3.3.2   Mobile Station GSM 850 and PCS 1900

Active Mode

The peak power measured in the conditions specified in clause A.3.1a, for a MS when allocated a channel, shall be no more than −36 dBm.

The peak power measured in the conditions specified in clause A.3.1b for a MS, when allocated a channel, shall be no more than:

- −36 dBm in the frequency band 9 kHz to 1 GHz;

- −30 dBm in all other frequency bands 1 GHz to 12.75 GHz.

The power emitted by the MS in a 100 kHz bandwidth using the measurement techniques for modulation and wide band noise (clause A.2.1) shall not exceed:

- −79 dBm in the frequency band 869 MHz to 894 MHz;
- −71 dBm in the frequency band 1930 MHz to 1990 MHz.

Idle Mode

The peak power measured in a 100 kHz bandwidth for a mobile, when not allocated a channel (idle mode), shall be no more than:

- −57 dBm in the frequency bands 9 kHz to 1000 MHz;
- −53 dBm in the frequency band 1850 MHz to 1910 MHz;
- −47 dBm in all other frequency bands 1 GHz to 12.75 GHz.

The power emitted by the MS in a 100 kHz bandwidth using the measurement techniques for modulation and wide band noise shall not exceed:

- −79 dBm in the frequency band 869 MHz to 894 MHz;
- −71 dBm in the frequency band 1930 MHz to 1990 MHz.

A maximum of five exceptions with a level up to −36 dBm are permitted in each of the band 869 MHz to 894 MHz and 1930 MHz to 1990 MHz for each ARFCN used in the measurements.

A.4 Radio Frequency Tolerance

The radio frequency tolerance for the base transceiver station and the MS is defined in GSM 05.10.

A.5 Output Level Dynamic Operation

NOTE: The term "any transmit band channel" is used here to mean:

- Any RF channel of 200 kHz bandwidth centered on a multiple of 200 kHz which is within the relevant transmit band.

A.5.1 Base Transceiver Station

The BTS shall be capable of not transmitting a burst in a timeslot not used by a logical channel or where DTX applies. The output power relative to time when sending a burst is shown in annex B. The reference level 0 dB corresponds to the output power level according to clause 4. In the case where the bursts in two (or several) consecutive timeslots are actually transmitted, at the same frequency, the template of annex B shall be respected during the useful part of each burst and at the beginning and the end of the series of consecutive bursts. The output power during the guard period between every two consecutive active timeslots shall not exceed the level allowed for the useful part of the first timeslot, or the level allowed for the useful part of the second timeslot plus 3 dB, whichever is the highest. The residual output power, if a timeslot is not activated, shall be maintained at, or below, a level of −30 dBc on the frequency channel in use. All emissions related to other frequency channels shall be in accordance with the wide band noise and spurious emissions requirements.

A measurement bandwidth of at least 300 kHz is assumed.

A.5.2 Mobile Station

The output power can be reduced by steps of 2 dB as listed in clause 4.1.

The transmitted power level relative to time when sending a burst is shown in annex B. The reference level 0 dB corresponds to the output power level according to clause 4. In the case of Multislot Configurations where the bursts in two or more consecutive timeslots are actually transmitted at the same frequency, the template of annex B shall be respected during the useful part of each burst and at the beginning and the end of the series of consecutive bursts. The output power during the guard period between every two consecutive active timeslots shall not exceed the level allowed for the useful part of the first timeslot, or the level allowed for the useful part of the second timeslot plus 3 dB, whichever is the highest. The timing of the transmitted burst is specified in GSM 05.10. Between the active bursts, the residual output power shall be maintained at, or below, the level of:

- −59 dBc or −54 dBm, whichever is the greater for GSM 400, GSM 900, and GSM 850, except for the timeslot preceding the active slot, for which the allowed level is −59 dBc or −36 dBm whichever is the greater;

- −48 dBc or −48 dBm, whichever is the greater for DCS 1800 and PCS 1900; in any transmit band channel.

A measurement bandwidth of at least 300 kHz is assumed.

The transmitter, when in idle mode, will respect the conditions of clause A.3.3.

A.6 Modulation Accuracy

A.6.1 GMSK Modulation

When transmitting a burst, the phase accuracy of the signal, relative to the theoretical modulated waveforms as specified in GSM 05.04, is specified in the following way.

For any 148-bits subsequence of the 511-bits pseudo-random sequence, defined in CCITT Recommendation O.153 fascicle IV.4, the phase error trajectory on the useful part of the burst (including tail bits), shall be measured by computing the difference between the phase of the transmitted waveform and the phase of the expected one. The RMS phase error (difference between the phase error trajectory and its linear regression on the active part of the time slot) shall not be greater than 5° with a maximum peak deviation during the useful part of the burst less than 20°.

NOTE: Using the encryption (ciphering mode) is an allowed means to generate the pseudorandom sequence. The burst timing of the modulated carrier in the active part of the time slot shall be chosen to ensure that all the modulating bits in the useful part of the burst (see GSM 05.04) influence the output phase in a time slot.

A.6.2 8-PSK Modulation

The modulation accuracy is defined by the error vector between the vector representing the actual transmitted signal and the vector representing the error-free modulated signal. The magnitude of the error vector is called error vector magnitude (EVM). For definition of the different measures of EVM, see annex G.

When transmitting a burst, the magnitude of the error vector of the signal, relative to the theoretical modulated waveforms as specified in GSM 05.04, is specified in the following way.

The magnitude of the error vector shall be computed by measuring the error vector between the vector representing the transmitted waveform and the vector representing the ideal one on the useful part of the burst (excluding tail symbols).

When measuring the error vector a receive filter at baseband shall be used, defined as a raised-cosine filter with roll-off 0.25 and single side-band 6 dB bandwidth 90 kHz.

The measurement filter is windowed by multiplying its impulse response by a raised cosine window given as:

$$w(t) = \begin{cases} 1, & 0 \leq |t| \leq 1.5T| \\ 0.5\big(1 + \cos\big[\pi(|t| - 1.5T)/2.25T\,\big]\big), & 1.5T \leq |t| \leq 3.75T \\ 0, & |t| \leq 3.75T \end{cases}$$

where T is the symbol interval.

The transmitted waveforms shall be normal bursts for 8-PSK as defined in GSM 05.02, with encrypted bits generated using consecutive bits from the 32767 bit length pseudorandom sequence defined in ITU-T Recommendation O.151 (1992).

A.6.2.1   RMS EVM

When transmitting a burst, the magnitude of the error vector of the signal, relative to the theoretical modulatedwaveforms as specified in GSM 05.04, is specified in the following way:

- The measured RMS EVM over the useful part of any burst, excluding tail bits, shall not exceed;

- For MS: under normal conditions 9.0% under extreme conditions 10.0%

- For BTS; after any active element and excluding the effect of any passive combining equipment:

- Under normal conditions 7.0% under extreme conditions 8.0%

- After any active element and including the effect of passive combining equipment: under normal conditions 8.0 % under extreme conditions 9.0%

The RMS EVM per burst is measured under the duration of at least 200 bursts.

A.6.2.2 Origin Offset Suppression

The origin offset suppression shall exceed 30 dB for MS and 35 dB for BTS under normal and extreme conditions.

A.6.2.3 Peak EVM

The peak value of EVM is the peak error deviation within a burst, measured at each symbol interval, averaged over at least 200 bursts to reflect the transient nature of the peak deviation. The bursts shall have a minimum distance in time of 7 idle timeslots between them. The peak EVM values are acquired during the useful part of the burst, excluding tail bits.

- The measured peak EVM values shall be 30% for MS and 22% for BTS under normal and extreme conditions. For BTS, the effect of any passive combining equipment is excluded.

A.6.2.4 95:th Percentile

The 95:th percentile is the point where 95% of the individual EVM values, measured at each symbol interval, is below that point. That is, only 5% of the symbols are allowed to have an EVM exceeding the 95:th-percentile point. The EVM values are acquired during the useful part of the burst, excluding tail bits, over 200 bursts.

The measured 95:th-percentile value shall be 15% for MS and 11% for BTS under normal and extreme conditions.

For BTS, the effect of any combining equipment is excluded.

A.7 Intermodulation Attenuation

The intermodulation attenuation is the ratio of the power level of the wanted signal to the power level of an intermodulation component. It is a measure of the capability of the transmitter to inhibit the generation of signals in its non-linear elements caused by the presence of the carrier and an interfering signal reaching the transmitter via the antenna.

A.7.1 Base Transceiver Station

An interfering CW signal shall be applied within the relevant BTS TX band at a frequency offset of 800 kHz, and with a power level 30 dB below the power level of the wanted signal.

The intermodulation products shall meet the requirements in clause A.7.2.

A.7.2 Intra BTS Intermodulation Attenuation

In a BTS intermodulation may be caused by combining several RF channels to feed a single antenna, or when operating them in the close vicinity of each other. The BTS shall be configured with each transmitter operating at the maximum allowed power, with a full complement of transceivers and with modulation applied. For the measurement in the transmit band the equipment shall be operated at equal and minimum carrier frequency spacing specified for the BSS configuration under test. For the measurement in the receive band the equipment shall be operated with such a channel configuration that at least 3rd order intermodulation products fall into the receive band.

A.7.2.1 GSM 400, GSM 900, DCS 1800

All the following requirements relate to frequency offsets from the uppermost and lowermost carriers. The peak hold value of intermodulation components over a timeslot, shall not exceed −70 dBc or −36 dBm, whichever is the higher, for frequency offsets between 6 MHz and the edge of the relevant Tx band measured in a 300 kHz bandwidth. 1 in 100 timeslots may fail this test by up to a level of 10 dB. For offsets between 600 kHz to 6 MHz the requirements and the measurement technique is that specified in clause A.2.1.

The other requirements of clause A.3.2 in the band 9 kHz to 12.75 GHz shall still be met.

A.7.2.2 MXM 850 and MXM 1900

All the following requirements relate to frequency offsets from the uppermost and lowermost carriers. The average value of intermodulation components over a timeslot, shall not exceed −60 dBc relative to the per carrier power, for frequency offsets 1.2 MHz to the edge of the relevant Tx band, measured in a 200 kHz bandwidth.

In addition to the requirements of this clause, the MXM 850 BTS and MXM 1900 BTS shall also comply with the applicable limits for spurious emissions established by the FCC rules for public mobile services [FCC Part 22, Subpart H] and FCC rules for wideband PCS services respectively.

NOTE: In some areas, to avoid uncoordinated system impacts, it may be beneficial to use a more stringent value.

- 70 dBc rms is suggested.

A.7.2.3 GSM 850 and PCS 1900

All the following requirements relate to frequency offsets from the upper-most and lowermost carriers. For frequency offsets 1.8 MHz to the edge of the relevant Tx band, measured in 300 kHz bandwidth the average value of intermodulation components over a timeslot shall not exceed −70 dBc relative to the per carrier power or −46 dBm, whichever is the higher. For offsets between 600 kHz and 1.8 MHz, the measurement technique and requirements are those specified in clause A.2.1, except for offsets between 1.2 MHz and 1.8 MHz, where the value of intermodulation components shall not exceed −70 dBc.

The other requirements of clause A.3.2 in the band 9 kHz to 12.75 GHz shall still be met.

In regions where additional protection between uncoordinated systems is required by national regulatory agencies, the intermodulation components for frequency offsets 1.2 MHz may be up to −60 dBc, if not violating the national regulation requirements. In this case the PCS 1900 BTS and GSM 850 shall also comply with the applicable limits for spurious emissions established by the FCC rules for wideband PCS services and FCC rules for public mobile services [FCC Part 22, Subpart H], respectively, or similar national requirements with comparable limits and rules.

A.7.3 Intermodulation Between MS (DCS 1800 & PCS 1900 Only)

The maximum level of any intermodulation product, when measured as peak hold in a 300 kHz bandwidth, shall be 50 dB below the wanted signal when an interfering CW signal is applied within the MS transmit band at a frequency offset of 800 kHz with a power level 40 dB below the power level of the wanted (DCS 1800 or PCS 1900 modulated) signal.

A.7.4 Mobile PBX (GSM 900 Only)

In a mobile PBX intermodulation may be caused when operating transmitters in the close vicinity of each other. The intermodulation specification for mobile PBXs (GSM 900 only) shall be that stated in clause A.7.2.

A.8 Receiver Characteristics

In this clause, the requirements are given in terms of power levels at the antenna connector of the receiver. Equipment with integral antenna may be taken into account by converting these power level requirements into field strength requirements, assuming a 0 dBi gain antenna. This means that the tests on equipment on integral antenna will consider field strengths (E) related to the power levels (P) specified, by the following formula (derived from the formula $E = P + 20\log F(\text{MHz}) + 77.2$):

- Assuming $F = 460$ MHz : $E\,(\text{dB}\mu\text{V/m}) = P\,(\text{dBm}) + 130.5$ for GSM 400;

- Assuming $F = 859$ MHz : $E\,(\text{dB}\mu\text{V/m}) = P\,(\text{dBm}) + 135.9$ for GSM 850;

- Assuming $F = 925$ MHz : $E\,(\text{dB}\mu\text{V/m}) = P\,(\text{dBm}) + 136.5$ for GSM 900;

- Assuming $F = 1795$ MHz : $E\,(\text{dB}\mu\text{V/m}) = P\,(\text{dBm}) + 142.3$ for DCS 1800;

- Assuming $F = 1920$ MHz : $E\,(\text{dB}\mu\text{V/m}) = P\,(\text{dBm}) + 142.9$ for PCS 1900.

Static propagation conditions are assumed in all cases, for both wanted and unwanted signals. For clauses A.1 and A.2, values given in dBm are indicative and calculated assuming a 50 ohms impedance.

A.8.1 Blocking Characteristics

The blocking characteristics of the receiver are specified separately for in-band and out-of-band performance as identified in the following tables.

Frequency Band	Frequency Range (MHz)			
	GSM 900		E-GSM 900	R-GSM 900
	MS	BTS	BTS	BTS
In-band	915–980	870–925	860–925	856–921
Out-of-band (a)	0.1–<915	0.1–<870	0.1–<860	0.1–<856
Out-of-band (b)	N/A	N/A	N/A	N/A
Out-of-band (c)	N/A	N/A	N/A	N/A
Out-of-band (d)	>980–12,750	>925–12,750	>925–12,750	>921–12,750

Frequency Band	Frequency Range (MHz)	
	DCS 1800	
	MS	BTS
In-band	1785–1920	1690–1805
Out-of-band (a)	0.1–1705	0.1–<1690
Out-of-band (b)	>1705–>1785	N/A
Out-of-band (c)	1920–1980	N/A
Out-of-band (d)	>1980–12,750	>1805–12,750

Frequency Band	Frequency Range (MHz)	
	PCS 1900 MS	PCS 1900 & MXM
In-band	1910–2010	1830–1930
Out-of-band (a)	0.1–<1830	0.1–<1830
Out-of-band (b)	1830–<1910	N/A
Out-of-band (c)	>2010–2070	N/A
Out-of-band (d)	>2070–12,750	>1930–12,750

Frequency Band	Frequency Range (MHz)	
	GSM 850 MS	GSM 850 & MXM 850 BTS
In-band	849–914	804–859
Out-of-band (a)	0.1–<849	0.1–<804
Out-of-band (b)	N/A	N/A
Out-of-band (c)	N/A	N/A
Out-of-band (d)	>914–12,750	>859–12,750

Frequency Band	Frequency Range (MHz)			
	GSM 450		GSM 480	
	MS	BTS	MS	BTS
In-band	457.6–473.6	444.4–460.4	486.0–502.0	472.8–<488.8
Out-of-band (a)	0.1–<457.6	0.1–<444.4	0.1–<486.0	0.1–<472.8
Out-of-band (b)	N/A	N/A	N/A	N/A
Out-of-band (c)	N/A	N/A	N/A	N/A
Out-of-band (d)	>473.6–12,750	460.4–12,750	>502.0–12,750	>488.8–12,750

The reference sensitivity performance as specified in Tables 1, 1a, 1b, 1c, 1d, and 1e shall be met when the following signals are simultaneously input to the receiver:

- For all cases except GSM 850 normal BTS, MXM 850 normal BTS, and MXM 1900 normal BTS, a useful signal, modulated with the relevant supported modulation (GMSK or 8-PSK), at frequency f_0, 3 dB above the reference sensitivity level or input level for reference performance, whichever applicable, as specified in clause 6.2;

- For GSM 850 normal BTS, MXM 850 normal BTS, and MXM 1900 normal BTS a useful signal, modulated with the relevant supported modulation (GMSK or 8-PSK), at frequency f_0, 1 dB above the reference sensitivity level or input level for reference performance, whichever applicable, as specified in clause 6.2;

- A continuous, static sine wave signal at a level as in the table below and at a frequency (f) which is an integer multiple of 200 kHz. For GSM 850 normal BTS, MXM 850 normal BTS, and MXM 1900 normal BTS at frequency offsets e 3000 kHz this signal is GMSK modulated by any 148-bit sequence of the 511-bit pseudorandom bit sequence, defined in CCITT Recommendation O.153 fascicle IV.4, with the following exceptions, called spurious response frequencies:

  - a) GSM 900 MS and BTS, GSM 850 MS and BTS, and MXM 850 BTS: in band, for a maximum of six occurrences (which if grouped shall not exceed three contiguous occurrences per group); DCS 1800, PCS 1900 MS and BTS and MXM 1900 BTS: in band, for a maximum of twelve occurrences (which if grouped shall not exceed three contiguous occurrences per group); GSM 400 MS and BTS: in band, for a maximum of three occurrences;

- b) out of band, for a maximum of 24 occurrences (which if below f_0 and grouped shall not exceed three contiguous occurrences per group). where the above performance shall be met when the continuous sine wave signal (f) is set to a level of 70 dBμV (emf) (i.e., -43 dBm).

Frequency Band	GSM 400, P-, E-, and R-GSM 900			DCS 1800 & PCS 1900			
	Other MS		Small MS	BTS	MS		BTS
	dBμV (emf)	dBm	dBμV (emf)	dBm	dBμV (emf)	dBm	dBμV (emf)	dBm
In-band 600 kHz ≤ |f–f0| < 800 kHz	75	−38	70	−43	87	−26	70	−43	78	−35
800 kHz ≤ |f–f0| < 1.6 MHz	80	−33	70	−43	97	−16	70	−43	88	−25
1.6 MHz ≤ |f–f0| < 3 MHz	90	−23	80	−33	97	−16	80	−33	88	−25
3 MHz ≤ |f–f0|	90	−23	90	−23	100	−13	−26	88	−25	87
Out-of-band (a)	113	0	113	0	121	8	113	0	113	0
Out-of-band (b)	—	—	—	—	—	—	—	—	101	−12
Out-of-band (c)	—	—	—	—	—	—	—	—	101	−12
Out-of-band (d)	113	0	113	0	121	8	113	0	113	0

NOTE: For definition of small MS, see clause 1.1.

The following exceptions to the level of the sine wave signal (f) in the above table shall apply:

For E-GSM MS, in the band 905 to 915 MHz	−5 dBm
For R-GSM 900 MS, in the band 880 MHz to 915 MHz	−5 dBm
For R-GSM 900 small MS, in the band 876 MHz to 915 MHz	−7 dBm
For GSM 450 small MS, in the band 450 MHz to 457.6 MHz	−5 dBm
For GSM 480 small MS, in the band 478.8 MHz to 486 MHz	−5 dBm
For GSM 900 and E-GSM 900 BTS, in the band 925 MHz to 935 MHz	0 dBm
For R-GSM 900 BTS at offsets 600 kHz abs(f–f0) MHz, in the band 876 MHz to 880 MHz	Level reduced by 5 dB

Frequency Band	GSM 850 MS		GSM 850 & MXM 850 BTS				MXM 1900 BTS	
	dBµV (emf)	dBm	dBµV (emf)	dBm			dBµV (emf)	dBm
In-band 600 kHz ≤ $|f-f_0|$ < 800 kHz	70	−43	76	−37			70	−43
800 kHz ≤ $|f-f_0|$ < 1.6 MHz	70	−43	78	−35			70	−38
1.6 MHz ≤ $|f-f_0|$ < 3 MHz	80	−33	80	−33			80	−33
3 MHz ≤ $|f-f_0|$	90	−23	80	−3			87	−33
Out-of-band (a)	113	0	121	8			113	0
(b)	—	—	—	—			101	—
(c)	—	—	—	—			101	—
(d)	113	0	121	8			113	0

The blocking characteristics of the micro-BTS receiver are specified for in-band and out-of-band performance. The ut-of-band blocking remains the same as a normal BTS and the in-band blocking performance shall be no worse than in the table below.

Frequency Band	GSM 900, GSM 850 and MXM 850 Micro- and Pico-BTS				DCS 1800, PCS 1900, and MXM 1900 Micro- and Pico-BTS			
	M1 (dBm)	M2 (dBm)	M3 (dBm)	P1 (dBm)	M1 (dBm)	M1 (dBm)	M1 (dBm)	P1 (dBm)
In-band 600kHz ≤$|f-f_0|$ <800kHz	−31	−26	−21	−34	−40	−35	−30	−41
800kHz ≤$|f-f_0|$ <1.6kHz	−21	−16	−11	−34	−30	−25	−20	−41
1.6kHz ≤$|f-f_0|$ <3MHz	−21	−16	−11	−26	−30	−25	−20	−31
3MHz ≤$|f-f_0|$	−21	−16	−11	−18	−30	−25	−20	−23

The blocking performance for the pico-BTS attempts, for the scenario of a close proximity uncoordinated MS, to balance the impact due to blocking by the MS with that due to wideband noise overlapping the wanted signal.

A.8.2   AM Suppression Characteristics

The reference sensitivity performance as specified in Tables 1, 1a, 1b, 1c, 1d, and 1e shall be met when the following signals are simultaneously input to the receiver.

- For all cases, except MXM 850 normal BTS and MXM 1900 normal BTS, a useful signal, modulated with the relevant supported modulation (GMSK or 8-PSK), at frequency f_0, 3 dB above the reference sensitivity level or input level for reference performance, whichever applicable, as specified in clause 6.2.

- For MXM 850 normal BTS and MXM 1900 normal BTS a useful signal, modulated with the relevant supported modulation (GMSK or 8-PSK), at frequency f_0, 1 dB above the reference sensitivity level or input level for reference performance, whichever applicable, as specified in clause 6.2.

- A single frequency (f), in the relevant receive band, | f–f$_0$ | 6 MHz, which is an integer multiple of 200 kHz, a GSM TDMA signal modulated in GMSK and by any 148-bit sequence of the 511-bit pseudo random bit sequence, defined in CCITT Recommendation O.153 fascicle IV.4, at a level as defined in the table below. The interferer shall have one timeslot active and the frequency shall be at least 2 channels separated from any identified spurious response. The transmitted bursts shall be synchronized to but delayed in time between 61 and 86 bit periods relative to the bursts of the wanted signal.

NOTE: When testing this requirement, a notch filter may be necessary to ensure that the cochannel performance of the receiver is not compromised.

	MS (dBm)	BTS (dBm)	Micro- and Pico-BTS			
			M1 (dBm)	M2 (dBm)	M3 (dBm)	P1 (dBm)
GSM 400	−31	−31	**	**	**	**
GSM 900	−31	−31	−34	−29	−24	−21
GSM 850	−31	−31	−34	−29	−24	−21
MXM 850	—	−33	−34	−29	−24	−21

Table (continued)

	MS (dBm)	BTS (dBm)	Micro- and Pico-BTS			
			M1 (dBm)	M2 (dBm)	M3 (dBm)	P1 (dBm)
DCS 1800	−29 / −31*	−35	−33	−28	−23	−26
PCS 1900	−29	−35	−33	−28	−23	−26
MXM 1900	—	−33	−33	−28	−23− −26	−26

NOTE 1: *The −31 dBm level shall only apply to DCS 1800 class 1 and class 2 MS meeting the −102 dBm reference sensitivity level requirement according to clause 6.2.

NOTE 2: **These BTS types are not defined.

A.8.3   Intermodulation Characteristics

The reference sensitivity performance as specified in tables 1, 1a, 1b, 1c, 1d and 1e shall be met when the following signals are simultaneously input to the receiver:

- A useful signal at frequency f_0, 3 dB above the reference sensitivity level or input level for reference

- Performance, whichever applicable, as specified in clause 6.2;

- A continuous, static sine wave signal at frequency $f1$ and a level of 70 dBμV (emf) (i.e., −43 dBm):

- For GSM 400 small MSs and GSM 900 small MSs and GSM 850 small MSs, DCS 1800 and PCS 1900 MS and DCS 1800, PCS 1900 and MXM 1900 BTS this value is relaxed to 64 dBμV (emf) (i.e., −49 dBm);

- For the DCS 1800 class 3 MS this value is relaxed to 68 dBμV (emf) (i.e., −45 dBm);

- Any 148-bits subsequence of the 511-bits pseudorandom sequence, defined in CCITT Recommendation O.153 fascicle IV.4 GMSK modulating a signal at frequency $f2$, and a level of 70 dBμV (emf) (i.e., −43 dBm):

- For GSM 400 small MSs and GSM 900 small MSs and GSM 850 small MSs, DCS 1800 and PCS 1900 MS and DCS 1800, PCS

1900 and MXM 1900 BTS this value is relaxed to 64 dBμV (emf) (i.e., −49 dBm);

- For the DCS 1800 class 3 MS this value is relaxed to 68 dBμV (emf) (i.e., −45 dBm); such that $f0 = 2f1 - f2$ and $|f2-f1| = 800$ kHz.

NOTE: For clauses 5.2 and 5.3 instead of any 148-bits subsequence of the 511-bits pseudorandom sequence, defined in CCITT Recommendation O.153 fascicle IV.4, it is also allowed to use a more random pseudorandom sequence.

A.8.4 Spurious Emissions

The spurious emissions for a BTS receiver, measured in the conditions specified in clause A.3.1, shall be no more than:

- 2 nW (−57 dBm) in the frequency band 9 kHz to 1 GHz;
- 20 nW (−47 dBm) in the frequency band 1 GHz to 12.75 GHz.

NOTE: For radiated spurious emissions for the BTS, the specifications currently only apply to the frequency band 30 MHz to 4 GHz. The specification and method of measurement outside this band are under consideration.

List of Acronyms

1G	first-generation wireless system
2G	second-generation wireless system
3G	third-generation wireless system
4G	fourth-generation wireless system
CAD	computer-aided design
CDMA	code division multiple access
CDPD	cellular digital packet data
CMOS	complementary metal-oxide semiconductor
CPW	coplanar waveguide
CVD	chemical vapor deposition
dB	decibel
DCS	dichlorosilane

DECT	Digital European Cordless Telecommunications
DRIE	deep reactive ion etching
EDP	diamine-pyrocatecol water
EM	electromagnetic
FBAR	film bulk acoustic resonators
GaAs	gallium arsenide
GHz	gigahertz
GSM	Global System for Mobile Communications
GPRS	general packet radio service
GPS	Global Positioning System
HF	hydrofluoric acid
IC	integrated circuit
IL	insertion loss
KHz	kilohertz
KOH	potassium hydroxide
LIGA	German acronym consisting of the letters LI (Roentgen-LIthographie, meaning X-ray lithography), G (Galvanik meaning electrodeposition), and A (Abformung meaning molding)
LNA	low-noise amplifier
mil	0.001 in
MEM	microelectromechanical

MHz	megahertz
MIC	microwave integrated circuit
MIM	metal-insulator-metal
MMIC	monolithic microwave integrated circuit
MUMPs	multiuser MEM processes
PCS	personal communications services
PECVD	plasma-enhanced chemical vapor deposition
RF	radio frequency
RIE	reactive ion etching
RFIC	radio-frequency integrated circuit
SEM	scanning electron micrograph
TEM	transverse electromagnetic
VCO	voltage-controlled oscillator
VSWR	voltage standing wave ratios
XeF$_2$	xenon difluoride

About the Author

Héctor J. De Los Santos was born in 1957 in Santo Domingo, Dominican Republic. He is principal scientist at Coventor in Irvine, California, where he leads Coventor's RF MEMS research and development effort. His activities include the conception, modeling, and design of novel RF MEMS devices. He received a Ph.D. from the School of Electrical Engineering, Purdue University, West Lafayette, Indiana, in 1989. From March 1989 to September 2000, he was employed at Hughes Space and Communications Company, Los Angeles, California, where he served as scientist and principal investigator and director of the Future Enabling Technologies IR&D Program. Under this program he pursued research in the areas of RF MEMS, quantum functional devices and circuits, and photonic band-gap devices and circuits. Dr. De Los Santos holds a dozen patents and has more than half a dozen patents pending. He is author of the textbook *Introduction to Microelectromechanical (MEM) Microwave Systems* (Artech House, 1999). His achievements are recognized in Maquis's *Who's Who in Science and Engineering,* Millennium Edition, and *Who's Who in the World,* 18th Edition. He is a senior member of the IEEE and member of Tau Beta Pi, Eta Kappa Nu, and Sigma Xi. He is an IEEE Distinguished Lecturer of the Microwave Theory and Techniques Society for the 2001–2003 term.

Index

Recent Titles in the Artech House Microelectromechanical Systems (MEMS) Series

An Introduction to Microelectromechanical Systems Engineering, Nadim Maluf

Introduction to Microelectromechanical (MEM) Microwave Systems, Héctor J. De Los Santos

RF MEMS Circuit Design for Wireless Communications, Héctor J. De Los Santos

For further information on these and other Artech House titles, including previously considered out-of-print books now available through our In-Print-Forever® (IPF®) program, contact:

Artech House
685 Canton Street
Norwood, MA 02062
Phone: 781-769-9750
Fax: 781-769-6334
e-mail: artech@artechhouse.com

Artech House
46 Gillingham Street
London SW1V 1AH UK
Phone: +44 (0)20 7596-8750
Fax: +44 (0)20 7630-0166
e-mail: artech-uk@artechhouse.com

Find us on the World Wide Web at:
www.artechhouse.com